COLLINS
GEM

BASIC FACTS
MATHEMATICS

Christopher Jones BSc
and
Peter Clamp BSc

Collins
London and Glasgow

Introduction

Basic Facts is a new generation of illustrated GEM dictionaries in important school subjects. They cover all the important ideas and topics in these subjects up to the level of first examinations.

Bold words in an entry means a word or idea developed further in a separate entry: *italic* words are highlighted for importance.

First published 1982
Reprint 10 9 8 7

© Wm. Collins Sons & Co. Ltd. 1982

ISBN 0 00 458889 4

Phototypeset and illustrated by
Parkway Illustrated Press

Printed in Great Britain by Collins, Glasgow

Symbols in Mathematics

Symbol	Meaning	Example
=	equals	$3 = 2 + 1$
+	add	$3 + 4 = 7$
−	subtract	$3 - 4 = -1$
×	multiply	$3 \times 4 = 12$
÷	divide	$3 \div 4 = 0.75$
$\sqrt{}$	square root	$\sqrt{9} = \pm 3$
n	power	$3^2 = 9$
$\angle$	angle	$\angle$ ABC
$\triangle$	triangle	$\triangle$ ABC
<	less than	$1 < 4$
≤	less than, or equal to	$4 \le 4$
>	greater than	$6 > 4$
≥	greater than, or equal to	$5 \ge 4$
≠	not equal to	$4.9 \ne 49$
≃	approximately equal to	$3.9 \simeq 4$
{ }	denotes a set	$\{1,2,3,...\}$
∈	is an element of	$1 \in \{1,2,3\}$
⊂	is a subset of	$A \subset B$
⊃	includes the subset	$B \supset A$
$\mathscr{E}$	universal set	$A \subset \mathscr{E}$
∅	empty set	$\{ \ \}$
∩	intersection	$A \cap B = \varnothing$
∪	union	$A \cup B = \mathscr{E}$

Abacus A frame containing **parallel** wires carrying beads, used for computation.

beads above bar are worth five times the beads below

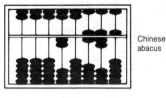

Chinese abacus

number shown at the bar is 20675

Abscissa The *x*-**coordinate**, or distance from the **vertical axis**, of a **point** referred to a system of **rectangular** coordinates.

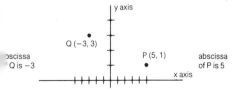

y axis

Q (−3, 3)

P (5, 1)

abscissa of Q is −3

abscissa of P is 5

x axis

1

Absolute The absolute **value** of a **real number** is its numerical value without regard for **sign**. The absolute value of the real number x (sometimes called **modulus** or mod x) is written $|x|$. For example: $|3.6| = 3.6$ $|-9.1| = 9.1$.

Acceleration is **rate** of change of **velocity** with time.

For example, if a body falls from rest with an acceleration of 9.8 **metres** per **second** per second, its velocity at intervals of a second, from 0 to 3 seconds, would be as shown in the table below:

time (seconds)	0	1	2	3
velocity (metres per second)	0	9.8	19.6	29.4

If **v** is a **vector** representing the velocity of a particle, then the acceleration of the particle is vector **a** obtained by **differentiating v** with respect to time

$$\mathbf{a} = \frac{d\mathbf{v}}{dt}$$

Acute An acute **angle** is an angle that measures between 0 and 90 **degrees**.

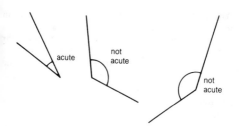

acute

not acute

not acute

Addition One of the basic **operations** of arithmetic. The **sum** of two **numbers** is determined by the operation addition, which is related to the process of accumulation, for example, $3 + 2 = 5$.

In abstract **algebra** used to denote certain operations applied to various **sets**, for example, in the set of 2×2 real **matrices**, addition is defined by:

$$\begin{pmatrix} a & b \\ c & d \end{pmatrix} + \begin{pmatrix} e & f \\ g & h \end{pmatrix} = \begin{pmatrix} a + e & b + f \\ c + g & d + h \end{pmatrix}$$

Affine An affine **transformation** of the **plane** is one that can be represented in the form

$$\begin{pmatrix} x \\ y \end{pmatrix} \rightarrow \begin{pmatrix} a & b \\ c & d \end{pmatrix} \begin{pmatrix} x \\ y \end{pmatrix} + \begin{pmatrix} e \\ f \end{pmatrix}$$

Examples include **translations**, **rotations**, **shears**, **reflections**, **stretches** and **enlargements**.

Under any affine transformation the **images** of sets of **parallel lines** are themselves sets of parallel lines.

3

Algebra The generalization, and representation in symbolic form, of significant results and patterns in **arithmetic** and other areas of mathematics, for example, $(4 + 3)(4 - 3) = 4^2 - 3^2$ and $(9 + 5)(9 - 5) = 9^2 - 5^2$ are both examples of the algebraic statement $(x + y)(x - y) = x^2 - y^2$

Algorithm A standard process designed to solve a particular set of problems.

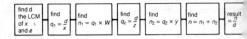

This is the familiar algorithm for the **addition** of fractions

$$\frac{w}{x} + \frac{y}{z}$$

Alternate Two **angles** are called alternate if they are on opposite sides of a **transversal** cutting two lines, and each has one of the lines for one of its sides.

In the diagram above a and c are an alternate pair of angles, as are b and d. Lines PQ and SR are **parallel** and in this case the alternate pairs of angles are equal $p = r$ and $q = s$.

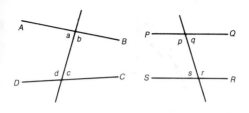

Altitude An altitude of a **triangle** is a **perpendicular** distance from a **vertex** of the triangle to the opposite side of the triangle, called the **base**.

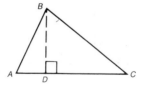

In this triangle AC is the base and the distance BD is the altitude.

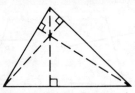

The **area** of any triangle can be calculated by
½ × base length × altitude.

Note: the three altitudes of any triangle pass
through a common point.

Amplitude In an oscillating motion the amplitude is a measure of the maximum **displacement**
from the **mean** or base position.

For the **periodic function** $f(x) = 3 \sin x$, the
amplitude is 3.

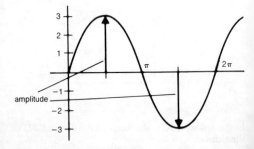

Analysis That branch of mathematics concerned with rigorous proofs of propositions mainly in the areas of **calculus** and **algebra**.

Angle The angle between two lines is a [...] of the inclination of the lines one to the other. Alternatively angle can be viewed as a measure [...] the **rotation** (about the point of **intersection** of t[...] lines) required to map one line onto the other.

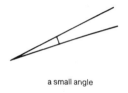

a small angle

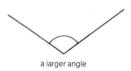

a larger angle

The two most common measures of angle are **degree** and **radian** measure.

7

Annulus An annulus is a **region** of a **plane** bounded by two **concentric circles**.

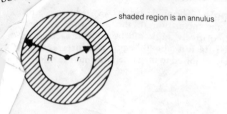

shaded region is an annulus

If the circles are of **radii** r and R, with $r < R$, then the **area** of the annulus is:

$$\pi R^2 - \pi r^2 \quad \text{or} \quad \pi(R + r)(R - r).$$

Antilogarithm The **inverse** of a **logarithm function**. Used particularly when base 10 logarithms are employed specifically for calculating purposes. For example:

$$\log_{10} 100 = 2 \qquad \text{antilog}_{10} 2 = 100$$

Apex The highest point of a **solid** or **plane** figure relative to a **base** plane or line.

apex

base

apex

base

apex

base

Approximation An approximation is an inexact result, which is accurate enough for some specific purpose.

For example, to the nearest tenth, the **square root** of 2 is 1.4. To the nearest hundredth it is 1.41.

The study of processes for approximating various forms in mathematics is an important branch of the subject.

For example, Taylor's Theorem provides a way of approximating to certain **functions** by **polynomials**.

Newton's Method gives an **iterative** way of obtaining approximations to the **roots** of certain **equations**.

9

Apse When a body moves in a central orbit (for example, an **ellipse**), any point on the orbit where the motion of the body is at **right angles** to the central **radius** vector is called an apse.

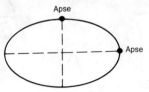

The distance from an apse point to the centre of the motion (apsidal distance) is a maximum or minimum value for the radius **vector**.

Arc An arc is a part of a curve.

In particular for the **circle**, the points A and B define both minor and major arcs of the circle.

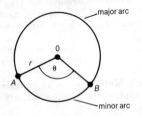

For the minor arc AB:

Arc length = $2\pi r \times \theta/360$, if θ is in **degrees**,
or
Arc length = $r\theta$, if θ is in **radians**.

Area The measure of the size of a surface. The areas of certain simple figures can be easily calculated.

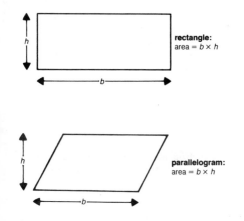

rectangle:
area = $b \times h$

parallelogram:
area = $b \times h$

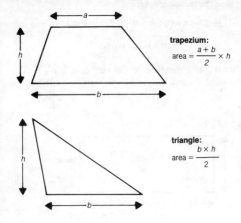

trapezium:

$$\text{area} = \frac{a + b}{2} \times h$$

triangle:

$$\text{area} = \frac{b \times h}{2}$$

Argand diagram A rectangular Cartesian grid used for diagrammatic representations of **complex numbers**.

The complex number 5 + 2i is represented by the point (5, 2).

The horizontal axis is called the real axis, and the vertical axis is called the imaginary axis.

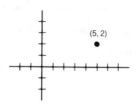

(5, 2)

Argument 1) The argument of a complex number is the angle between the position vector of the number on the Argand diagram, and the positive real axis. In the diagram above, the argument of the complex number $5 + 2i$ is $21.8°$.

2) the **independent** variable of a function is sometimes called the argument of the function. For example:

x is the argument of the function $y = 3x^2 + 6x - 7$.

n is the argument of the function $A = 2n - 4$.

f is the argument of the function $C = \dfrac{(f - 32) \times 5}{9}$

Arithmetic The study of number, particularly with regard to simple operations: addition, subtraction, division and multiplication, and their application to solutions of problems.

1) Arithmetic mean Commonly referred to as the average of a given set of numbers. The arithmetic mean of n numbers $a_1\, a_2 \ldots a_n$ is calculated by $(a_1 + a_2 + a_3 + \ldots a_n) \div n$.

For example, the arithmetic mean of 5, 7, 1, 8, 4 is $(5 + 7 + 1 + 8 + 4) \div 5 = 5$.

(2) Arithmetic progression A sequence in which there is a common difference between any member of the sequence and its successor.

For example, 3, 7, 11, 15, 19, 23 is an arithmetic progression with common difference 4.

Associative Any operation $*$ which has the property $a * (b * c) = (a * b) * c$ for all members a, b and c of a given set, is called associative.

Multiplication of real numbers is associative, for example, $8 \times (5 \times 2) = (8 \times 5) \times 2$.

Division is not associative, for example, $80 \div (4 \div 2) \neq (80 \div 4) \div 2$

Asymptote A line approached, but never reached, by a curve.

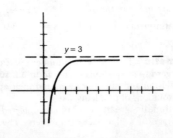

14

The curve of the graph $y = 3 - 1/x$ has the line $y = 3$ as an asymptote. The curve can be drawn as close to $y = 3$ as desired by taking sufficiently large values of x.

Average   An average is a single number that represents or typifies a collection of numbers. **Mode**, **mean** and **median** are three commonly used averages.

Axiom   A principle, taken to be self evident, and not requiring proof.

'Things equal to the same thing are equal to each other' is one of the axioms attributed to Euclid.

Axis (plural, axes).   An axis is a significant line which a graph, plane shape or solid can be referenced to. For example, **coordinate** axes, axes of **symmetry** etc

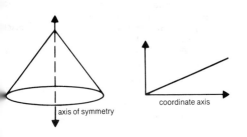

axis of symmetry

coordinate axis

Bar chart A graph using parallel bars to illustrate information. The lengths of the bars are proportional to the quantity represented. Often used to illustrate statistical data.

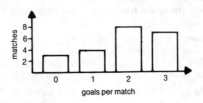

Base 1) Number base is the method of grouping used in a number system that relies on place value. In the decimal number system grouping and place value is based on 10. 423 means

$3 + (2 \times 10) + (4 \times 100)$. In base 5, 423 would mean $3 + (2 \times 5) + (4 \times 25)$.

2) In expressions like 4^5, 4 is called the base whilst 5 is called the **exponent**. This use of the term base is closely linked to its use in the context of logarithms.

In base 10 logarithms log 47 = 1.6721, this means that $10^{1.6721} = 47$.

Bearing is a navigational and surveying term. The bearing of a point A from an observer O is the angle between the line OA and the north line through O, measured in a clockwise direction.

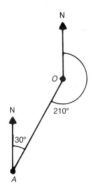

In the diagram the bearing of A from O is 210°. The three figure bearing of O from A is 030°.

Billion A million millions (1 000 000 000 000). In the U.S.A. it means a thousand millions.

Bimodal   A term used to describe distributions of data that show two **modes** or peaks in frequency.

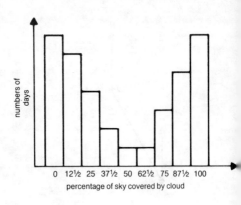

percentage of sky covered by cloud

Binary   A number system based on grouping i twos. Binary notation requires the use of just tw symbols 0 and 1, and its column headings or plac values are: . . . 64   32   16   8   4   2   1. Therefo

twenty nine is: 1   1   1   0   1   in binary
ninety two is: 1   0   1   1   1   0   0   in binary

Binomial A **polynomial** in two variables, for example, $7x + 4y$.

The numbers in the rows of **Pascal's triangle** are called binomial coefficients.

$$\begin{array}{ccccccccccc}
 & & & & & 1 & & & & & \\
 & & & & 1 & & 1 & & & & \\
 & & & 1 & & 2 & & 1 & & & \\
 & & 1 & & 3 & & 3 & & 1 & & \\
 & 1 & & 4 & & 6 & & 4 & & 1 & \\
1 & & 5 & & 10 & & 10 & & 5 & & 1
\end{array}$$

The numbers in successive rows correspond to the coefficients of the xy terms when $(x + y)^n$ is expanded for various values of n. For example:

$$(x + y)^5 = x^5 + 5x^4y + 10x^3y^2 + 10x^2y^3 + 5xy^4 + y^5$$

Bisect To bisect is to cut in half. The term is often used in a geometrical context.

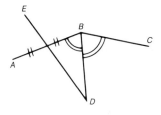

The line BD bisects $\angle ABC$. The line DE bisects the line AB.

19

Brackets Brackets are used to indicate the order in which operations are to be carried out. For example:

$$7 + (2 \times 3) = 7 + 6 = 13 \qquad (7 + 2) \times 3 = 9 \times 3 = 27$$

Elementary algebra is often concerned with the introduction or removal of brackets in cases where one operation **distributes** over another. For example:

$$6x^2y + 9xy^3 = 3xy\,(2x + 3y^2)$$

Calculate To perform a mathematical process often numerical, to obtain some desired result.

Calculus An important branch of mathematics concerned with the study of the behaviour of functions.

Differential calculus is concerned with the variation of functions, with maximum and minimum values, gradients, approximations to functions etc.

Integral calculus is concerned with calculations of areas and volumes, problems of summation etc.

The techniques of calculus were developed independently by Newton and Leibnitz in the 17th century. Calculus has important applications in many areas of the natural and physical sciences.

Cancellation The process of producing, from a given fraction, an **equivalent** fraction, by dividing **numerator** and **denominator** by a common factor. For example:

$$\frac{15}{20} = \frac{3 \times 5}{4 \times 5} = \frac{3}{4}$$

Cardinal number That aspect of a number that reflects the manyness of elements in a set. The cardinal number 3 is the property shared by all sets containing three elements.

Cardioid The **locus** of a point on the circumference of a circle, which rolls on a fixed circle of equal radius.

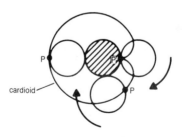

cardioid

Cartesian coordinates   A method of locating the position of a point on a plane or in space.

In the case of a plane the distances of the point from two perpendicular axes, given as an ordered pair of real numbers (x, y).

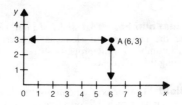

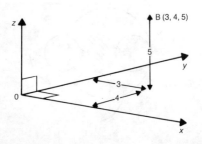

The cartesian coordinates of the point A are (6, 3).

In space, the distances of the point from three mutually perpendicular planes, which in turn form three mutually perpendicular axes, as they intersect in pairs. The distances given as an ordered triple of real numbers (x, y, z).

The Cartesian coordinates of the point B are (3, 4, 5).

Catenary The shape of the curve assumed by a heavy, flexible cable or rope, when suspended from two points.

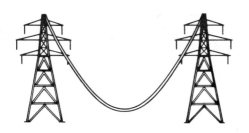

Centigrade The unit of temperature in which the freezing and boiling temperatures of water are assigned 0 degrees and 100 degrees respectively.

37 degrees centigrade is written 37°C

Centimetre A centimetre is one-hundredth of a **metre**.

A penny is two centimetres in diameter. This is written 2 cm.

Characteristic When using common, base 10 logarithms, the characteristic is the integer part of the logarithm when written in the conventional way. The characteristic conveys information regarding the position of the decimal point in the original number. For example:

$0.041 = 10^{-2} \times 4.1$
$\log (0.041) = \log (10^{-2} \times 4.1)$
$\qquad\quad = \log (10^{-2}) + \log (4.1)$
$\qquad\quad = -2 + 0.6128$
$\qquad\quad$ written $\bar{2}.6128$

$\bar{2}$ is the characteristic of the logarithm.

Chord   A line joining two points on a curve.

examples of chords

25

Circle A plane curve formed by the set of all points at a given fixed distance from a fixed point. The fixed point is called the centre, and the distance the radius.

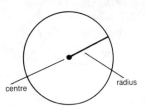

centre

radius

Circles have line and rotational symmetry of infinite order about the centre.

For a fixed perimeter, a circle encloses the maximum area for any plane curve.

Circumference Used to mean both the boundary line of a circle and the length of the boundary line, or distance round the circle.

The ratio circumference/diameter is a constant for all circles, and is denoted by the Greek letter π.

Numericallly π is an irrational number of approximate value 3.14 (to three significant figures). The circumference of a circle may therefore be calculated using the formulae: circumference = π × diameter. For example, for a circle of diameter 6 cm: circumference $\simeq$ 3.14 × 6 cm ($\simeq$ 18.8 cm).

Class interval A grouping of statistical data to enable the data to be represented and interpreted in a

29, 33, 49, 71, 16, 53, 62, 81,
39, 46, 52, 42, 31, 26, 11, 62,
69, 26, 60, 21, 44, 19, 79, 50,
43, 48, 55, 27, 12, 42, 70, 74,
29, 33, 60, 53, 48, 31, 15, 88,
31, 26, 77, 21, 39, 54, 40, 70,
45, 62.

simpler way. For example, the fifty examination percentages could be grouped into the following class intervals:

Mark	1 – 25	26 – 50	51 – 75	76 – 100
Frequency	7	24	15	4

Clockwise An indication of direction of rotation in a plane. A clockwise rotation has the same sense as the rotation of the hands of a clock.

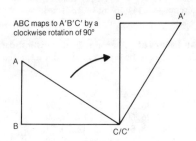

ABC maps to A'B'C' by a clockwise rotation of 90°

Closed 1) A set is closed under an operation when the result of combining together any two members of the set, using the operation, always results in a member of the original set.

For example, the set of Natural numbers {1, 2, 3, 4, . . .} is closed under the operations of addition and multiplication, but is not closed under division or subtraction:

$3 \div 2 = 1.5$, which is not a natural number
$1 - 4 = -3$, which is not a natural number

2) A closed curve is one with no end points.

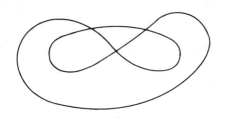

3) A closed interval on the real number line is a set of all numbers x, defined by inequalities of the type $a \leq x \leq b$.

The closed interval $\{x: 1 \leq x \leq 3\}$ is the set of all real numbers between, and including, the 'end-points' 1 and 3.

Coefficient 1) In simple algebra the coefficient of a term is the numerical part of the term. For example, in $4x^2y + 3xy^2 - 12x + 9y$

the coefficient of x^2y is 4
the coefficient of xy^2 is 3
the coefficient of x is -12
the coefficient of y is 9

2) In science and engineering the term is often used to denote a specific numerical constant that is characteristic of a particular system, for example, coefficient of linear expansion, coefficient of friction, etc.

Collinear Points are collinear if they lie in a straight line.

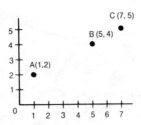

The three points A, B and C are collinear.

Notice the vectors **AB** and **BC** are simply connected: **AB** = 2**BC**.

Notice also that there is a connection between the position vectors of A, B and C:
OB = ⅓**OA** + ⅔**OC**.

Column A vertical array of elements.

The **matrix** $\begin{pmatrix} 6 & 1 & 3 \\ 4 & 2 & 7 \end{pmatrix}$ has three columns.

Combination A combination of a set of n objects is any selection of n or less objects from the set, regardless of order.

For example, the full list of combinations of three letters from the set $\{a, b, c, d\}$ is: abc, abd, acd, bcd.

The number of combinations of r objects that can be made from a set of n objects is usually denoted by:

$$_nC_r \quad \text{or} \quad \binom{n}{r}$$

It can be shown that:

$$_nC_r = \frac{n!}{r!(n-r)!} \quad \text{(where } n! \text{ means \textbf{factorial} } n\text{)}$$

For example, $\qquad _{12}C_5 = \dfrac{12!}{5! \times 7!} = 792.$

Notice that $_{12}C_5 = {}_{12}C_7$ and in general $_nC_r = {}_nC_{n-r}$.

The values of $_nC_r$ for varying n and r can be found from the appropriate row of **Pascal's Triangle**.

Common denominator In preparation for the addition or subtraction of fractions with unlike denominators, the fractions are usually converted into equivalent fractions with the same, or common, denominator.

$$\frac{1}{3} + \frac{1}{2} = \frac{2}{6} + \frac{3}{6} \qquad \text{6 is the common denominator}$$

Common difference The difference between successive terms in an arithmetic progression. For example:

    8, 11, 14, 17, 20   common difference 3
    19, 17, 15, 13, 11   common difference −2

Common logarithm Logarithms to the base of ten are called common logarithms.

Commutative Any operation * which has the property a * b = b * a for all members a and b of a given set, is called commutative.

    Multiplication of real numbers is commutative, for example 7 × 3 = 3 × 7.

    Subtraction is not commutative, for example, 7 − 3 ≠ 3 − 7

Complement The complement of a set S is the set consisting of the elements within the **universal** set that are not in S

    The complement of S is usually denoted by S′.

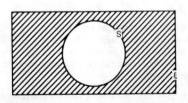

32

The shaded portion of the diagram represents S'.

if $\mathscr{E}$ = {letters of the alphabet}
and c = {consonants}
then c' = {vowels}

Complementary angles Two angles whose sum is 90° are complementary angles.

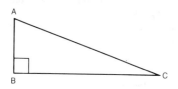

In the right-angled triangle ABC, $\angle$BAC and $\angle$BCA are complementary angles.

   $\angle$BAC is the complement of $\angle$BCA.

Complete the square A method for solving quadratic equations by applying suitable transformations to the equation to reduce it to the form:

$(x + h)^2 = k$

For example:
$x^2 + 12x - 3 = 0 \Rightarrow x^2 + 6x - 1.5 = 0$
$\Rightarrow x^2 + 6x = 1.5 \Rightarrow (x + 3)^2 - 9 = 1.5$
$\Rightarrow (x + 3)^2 = 10.5 \Rightarrow x + 3 = \pm 3.24$
$\Rightarrow x = 0.24$ or $x = -6.24$

Complex numbers An extension of the real number system designed originally to overcome difficulties involving the solution of problems like $x^2 + 3 = 0$, which have no solution in the **real numbers**.

Traditionally complex numbers are written in the form $a + bi$, where a and b are real numbers, and $i^2 = -1$: a and b are called respectively the real and imaginary parts.

To add: $(a + bi) + (c + di) = (a + c) + (b + d)i$.
To multiply:
$(a + bi) \times (c + di) = (ac - bd) + (ad + bc)i$.

More formally, complex numbers can be viewed as pairs of real numbers (a, b) with a particular multiplication and addition defined, so that numbers of the form $(a, 0)$ behave as real numbers, and $(0, 1)$ behaves as $\sqrt{-1}$.

Complex numbers have important implications in many branches of pure and applied mathematics.

Component A component of a **vector** is one of a set of two or more often mutually perpendicular vectors that are equivalent in effect to the given vector.

If vector **OA** is of **magnitude** 5 and at $30°$ to the x-axis then:

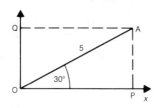

Component in the direction of x-axis is vector **OP**.
Magnitude = 5 cos 30° = 4.33.

Component in the direction of y axis is vector **OQ**.
Magnitude = 5 sin 30° = 2.5.

Composite function The function $f: x \rightarrow (3x + 1)^2$,
can be thought of as the composition of two simpler
functions:

$$g: x \rightarrow (3x + 1) \text{ and } h: x \rightarrow x^2$$
$$x \overset{g}{\rightarrow} (3x + 1) \overset{h}{\rightarrow} (3x + 1)^2$$
$$f = hg$$

Notice the notation: $f = hg$ means f is equivalent to the
function g followed by h.

Any function that can be split into two or more
simpler functions in this way is called a composite
function. For example:

if $f: x \rightarrow \sin^2 4x$, then $f = pqr$, where $r: x \rightarrow 4x$,
$q: x \rightarrow \sin x$ and $p: x \rightarrow x^2$.

35

Compound interest A method of calculating **interest** on money where the interest earned during a period is calculated on the basis of the original sum together with any interest earned in previous periods.

£100 invested at 10 per cent per year for three years would appreciate in the following way:

Original sum → end of first year → end of second year → end of third year

Original sum	end of first year $£100 \times 1.1$	end of second year $£110 \times 1.1$	end of third year $£121 \times 1.1$
£100	£110	£121	£133.10

In general if £P is invested at r per cent compound interest, then after n years it would be worth:

$$£P \times \left(\frac{100 + r}{100} \right)^n$$

Computer An electronic device designed to process large ammounts of coded information rapidly.

Computer systems commonly comprise three main elements: *Input devices,* for example, keyboard paper tape etc., *central processor, output devices,* for example, television monitor, printer, etc.

Rapid developments in microelectronic technology have resulted in drastic reductions in the size and cost of computers, whilst increasing their processing capabilities.

Concave A concave curve or surface is hollow towards a given point of reference.

36

The dish of a radio telescope presents a concave surface towards the sky.

Cone A cone is a solid bounded by a plane (often circular) base, and tapering to a fixed point called the vertex. A right cone is one in which the axis, or line joining the centre of symmetry of the base to the vertex, is perpendicular to the base.

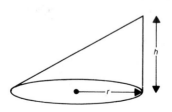

Volume of cone is:

$$V = \tfrac{1}{3}\pi r^2 h$$

where r = base radius, and h = height.

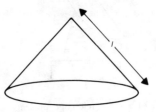

Curved surface area of a right cone is:

$$A = \pi r l$$

where l is the slant height.

Congruent Shapes, plane or solid, are called congruent if they have the same shape and size.

Certain simple transformations such as rotations, reflections and translations map objects onto congruent images.

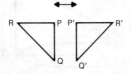

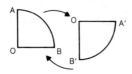

Conic A curve obtained by taking a plane section through a cone.

By varying the angle of cut, four main types of conic can be obtained.

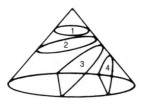

Alternatively a conic can be thought of as the **locus** of a point that moves so that the ratio of its distance from a fixed point to its distance from a fixed line is constant. This ratio is called the **eccentricity** of the conic.

Conjugate Two angles whose sum is 360° are called conjugate.

The conjugate of the **complex number** $a + bi$ is $a - bi$.

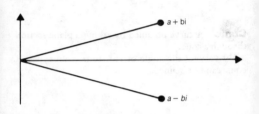

A complex number and its conjugate are related by a reflection in the real axis of the **Argand diagram**.

The product of a complex number and its conjugate is equal to the square of its **modulus**

$$(a + bi) \times (a - bi) = a^2 + b^2.$$

Constant A fixed quantity in an expression is called a constant.

In the expression $y = 3x + 2$, 3 and 2 are constants whilst x and y are variables.

Construction Additional points or lines to supplement a geometrical figure in order to prove some property of the figure are called constructions.

Continuity Quantities like the number of pupils in a class can be measured precisely, and vary in integer steps.

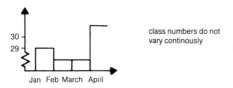

class numbers do not vary continuously

When measuring height, however, the result can never be exact. In growing from 160 cm to 161 cm we assume every possible value between 160 and 161 has been attained.

Quantities such as length, weight, temperature, speed, time, etc., are continuous variables.

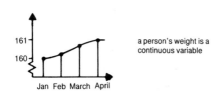

a person's weight is a continuous variable

Converse The converse of the statement 'if a number ends in zero it is divisible by ten' is 'if a number is divisible by ten, then it ends in zero'.

Many important statements in mathematics are in the form 'if x then y'. The converse of such a statement is 'if y then x'.

The converse of a true statement need not itself be true. For example 'if p and q are even, then $p + q$ is even' is a true statement. The converse 'if $p + q$ is even, then p and q are even' is false.

Conversion The process involved in changing from one system (often of units) to another.

For example, to express a temperature, given in the **Fahrenheit** scale, in **Centigrade**, use the conversion formula:

$$C = \frac{5}{9} \ (F - 32)$$

Convex A curve or surfaces that bulges towards a given point of reference is called convex.

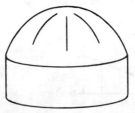

The dome of St. Paul's Cathedral presents a convex surface to the sky.

A convex figure is one in which any line joining two points on the figure is contained within the figure.

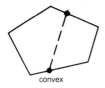

convex

not convex

Coordinates A coordinate system is used to locate points on a plane or in space. **Cartesian** and **polar** are two commonly used coordinate systems.

Coplanar Sets of points or lines lying in the same plane are called coplanar.

Correlation A statistical term used to describe a relationship between two varying quantities, when changes in one variable are linked to changes in the other. A positive correlation occurs when increases and decreases in the two variables happen together. A negative correlation occurs when increases in one variable are associated with decreases in the other.

Doctors, for example, have suggested a positive correlation between the habit of cigarette smoking and the incidence of heart disease.

A correlation coefficient is a number between -1 and 1, that gives a measure of correlation between the values of two variables.

Correspondence A relationship involving the matching of pairs of elements from two sets.

Four types of correspondence are commonly identified:

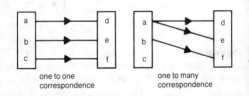

one to one
correspondence

one to many
correspondence

44

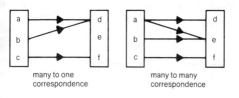

many to one
correspondence

many to many
correspondence

Corresponding points, lines and angles

When two figures are related by a simple transformation. Features of the first figure are said to correspond to their images on the transformed figure.

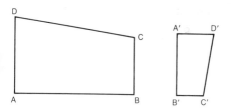

ABCD and A'B'C'D' are **similar** rectangles, C and C' are corresponding points, ∠DAB and ∠D'A'B' are corresponding angles.

When a transversal cuts a pair of lines, as in the diagram below, the pairs of angles p and p′, q and q′, r and r′, s and s′ are called corresponding angles.

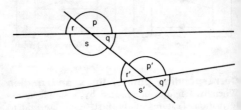

If the pair of lines cut by the transversal are parallel, the corresponding angles are equal.

Cosecant A trigonometric function.

In a right-angled triangle, the cosecant of an angle is the ratio hypotenuse/opposite side.

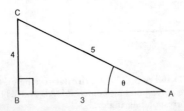

In the triangle ABC above:

$$\csc \theta = \frac{AC}{BC} = \frac{5}{4} = 1.25$$

The value of θ can now be determined by examining a table of values for the cosecant function $\theta = 53.1°$.

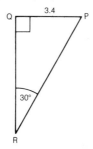

In the triangle PQR above:

$$\csc 30° = \frac{RP}{QP}$$

But from the table of values for the cosecant function $\csc 30° = 2$.

Hence $2 = \dfrac{RP}{3.4}$ and $RP = 6.8.$

Cosine   A trigonometric function.

In a right-angled triangle the cosine of an angle is the ratio adjacent side/hypotenuse.

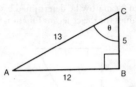

In the triangle ABC above:

$$\cos \theta = \frac{CB}{CA} = \frac{5}{13} = 0.385$$

The value of θ can now be determined by examining a table of values for the cosine function. θ = 67.4°.

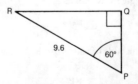

In the triangle PQR above:

$$\cos 60° = \frac{PQ}{PR}$$

But from the table of values for the cosine function, cos 60° = 0.5.

Hence $0.5 = \frac{PQ}{9.6}$ and PQ = 4.8.

Cotangent A trigonometric function.

In a right-angled triangle the cotangent of an angle is the ratio adjacent side/opposite side.

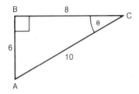

In the triangle ABC above:

$$\cot \theta = \frac{BC}{BA} = \frac{8}{6} = 1.33$$

The value of θ can now be determined by examining a table of values for the cotangent function. θ = 36.9.

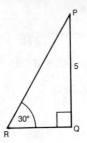

In the triangle PQR above:

$$\cot 30° = \frac{QR}{QP}$$

but from the table of values for the contangen function $\cot 30° = 1.73$.

Hence $1.73 = \dfrac{QR}{5}$ and $QR = 8.65$.

Count To count the number of objects in a set to match the objects in a one-to-one fashion with th names of the positive integers – 'one, two, thre . . . .'

The positive integers or natural numbers are ofte called counting numbers.

Cross multiply A term used to describe a pr cess of simplifying equations involving fraction erms. For example:

If $\dfrac{4x}{3} = 6$, then $x = \dfrac{6 \times 3}{4}$ so $x = 4\frac{1}{2}$.

Cross section The cross section of a solid is the plane shape that results from cutting through the solid, often at right angles to an axis of symmetry of the solid.

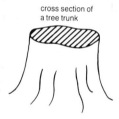

cross section of
a tree trunk

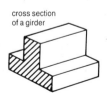

cross section
of a girder

Cube A regular solid with six square faces, having all its edges equal in length.

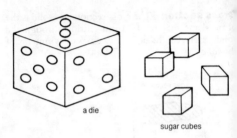

a die

sugar cubes

To cube a number is to raise the number to th power of three. 'Four cubed' is written
$4^3 = 4 \times 4 \times 4 = 64$.

Cubic A cubic function is a **polynomial** of th third degree:

$$ax^3 + bx^2 + cx + d$$

All cubic equations with real coefficients have least one real solution.

The cubic metre, cubic centimetre, etc., are uni used in measuring volume. A solid with volum three cubic metres, written 3 m^3, has the sam volume as three cubes each with 1 m edges.

Cuboid A cuboid is a solid with six rectangular faces, the opposite faces being equal in size.

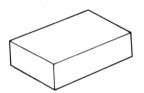

Cumulative frequency In statistics a method of grouping the frequencies of the value of some variable by adding the frequencies not greater than certain values of the variable.

Frequency table for pupils' marks in a test

Cumulative frequency table for same marks.

mark	frequency
0	4
1	5
2	8
3	12
4	10
5	3

mark	cumulative frequency
≤0	4
≤1	9
≤2	17
≤3	29
≤4	39
≤5	42

Cusp   A simple cusp is a point on a curve where the curve crosses itself, and where the two branches of the curve share a common tangent, the branches being on opposite sides of the **tangent**. For example, $y^2 = x^3$.

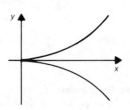

The curve above has a simple cusp at the origin, where the x-axis is the common tangent.

Cyclic   A cyclic **permutation** of an ordered set is achieved by mapping each object to its successor, the last object being mapped to the first. For example:

$$(1, 2, 3) \rightarrow (2, 3, 1) \rightarrow (3, 1, 2)$$

The first permutation is denoted by:

$$\begin{pmatrix} 1 & 2 & 3 \\ 2 & 3 & 1 \end{pmatrix}$$

A cyclic **quadrilateral** is a four-sided plane shape having vertices lying on the perimeter of a circle.

The opposite angles of a cyclic quadrilateral are **supplementary**.

Cycloid A cycloid is the curve traced by a point P on the circumference of a circle which rolls along a straight line.

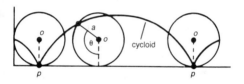

For a circle of radius a the **parametric** equation of a cycloid is:

$$x = a(\theta - \sin \theta) \quad y = a(1 - \cos \theta)$$

Cylinder A cylinder is a solid with one axis of symmetry about which it has a uniform circular cross section.

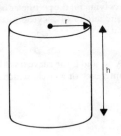

The volume of a cylinder is given by:

$V = \pi r^2 h$ (The area of a circular end multiplied by the height.)

The curved surface area of a cylinder is given by:

$A = 2\pi rh$ (The **circumference** of a circular end multiplied by the height.)

Data Information often in numerical form, which has been collected for statistical purposes.

Decagon A decagon is a **polygon** with ten sides.

Decimal A general term applied to the system of representing whole and fractional numbers in the base of ten.

A decimal fraction is one with denominator being a power of ten, for example, $^3/_{10}$, $^7/_{100}$, $^{38}/_{1000}$, etc.

The base 10 place value system for whole numbers is extended in the decimal system to include the representation of decimal fractions.

$$13 \frac{17}{100} = 13.17$$

tens   tenths
units   hundredths

The decimal point is used to identify place values as indicated above.

All **rational numbers** can be represented in decimal form by a **terminating** or **recurring** sequence of decimal digits.

Degree   a) A unit for measuring angles, in which one complete rotation is divided into 360 degrees (written 360°).

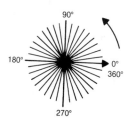

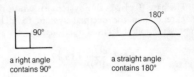

a right angle
contains 90°

a straight angle
contains 180°

b) The degree of a **polynomial** expression or equation of one variable is the highest power to which the unknown is raised. For example:

$$7x^5 - 5x^3 + 19x^2 - 3 \text{ is of degree 5 in } x.$$

An expression in several variables, for example,

$$8x^3y^2z^4$$

is said to be of degree 9 (the sum of the powers of x, y and z), but of degree 2 in y.

Denominator   The 'bottom number' of a fraction. For example, the denominator of ¾ is 4.

Depression   The angle of depression of an object from an observer viewing the object from above, is the angle between the line joining the object and the observer, and the horizontal plane.

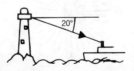

The angle of depression from the lighthouse to the ship is 20°.

Derivative If $y = f(x)$ is a **function** of the **variable** x, the derivative of $y = f(x)$ at some $x = a$, is the rate of change of the function with respect to x at a.

The derivative is often interpreted as the slope of the graph of $y = f(x)$ at $x = a$.

The derivative of $y = f(x)$ is usually written dy/dx or $f'(x)$. For $f(x) = x^3$ $f'(x) = 3x^2$, thus the rate of change of x^3 at $x = 4$ is $3 \times 4^2 = 48$.

Determinant The determinant of a square matrix M is a number associated with that matrix that provides various information concerning the nature of the matrix.

For the 2×2 matrix $M = \begin{pmatrix} a & b \\ c & d \end{pmatrix}$ it is written

$$\begin{vmatrix} a & b \\ c & d \end{vmatrix} \quad \text{or Det M, and is equal to } ad - bc.$$

The determinants of larger matrices can be evaluated in a similar way. For example:

$$\begin{vmatrix} a & b & c \\ d & e & f \\ g & h & i \end{vmatrix} = a \begin{vmatrix} e & f \\ h & i \end{vmatrix} - b \begin{vmatrix} d & f \\ g & i \end{vmatrix} + c \begin{vmatrix} d & e \\ g & h \end{vmatrix}$$

The theory of determinants is particularly useful in the application of matrices to solving sets of linear equations, and linear transformations.

Diagonal A diagonal of a **polygon** is a line joining any two non-adjacent **vertices**.

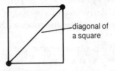

diagonal of a square

An n sided polygon has $n - 3$ diagonals from any vertex, and $\frac{1}{2}n(n - 3)$ distinct diagonals in total.

Diameter The diameter of a circle is a **chord** that passes through the centre of the circle.

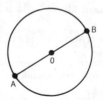

If 0 is the centre of the circle then AB is a diameter of the circle.

The length of the diameter of a circle is twice the length of the **radius** of the circle.

Difference The difference between two numbers a and b is the result obtained by subtracting one from the other, i.e., $a - b$.

The difference of two squares $a^2 - b^2$ is the common name given to the factorization
$$a^2 - b^2 = (a - b)(a + b).$$

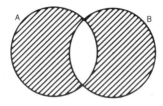

The symmetric difference of two sets A and B is $(A \cap B') \cup (B \cap A')$ or $(A \cup B) \cap (A \cap B)'$ and is represented by the shaded area on the **venn diagram** above.

Differentiation The process of finding the **derivative** $f'(x)$ of a function f.

Essentially the process involves the evaluation of:

$$\text{Limit } \frac{f(x + h) - f(x)}{h} \text{ as } h \text{ tends to } 0.$$

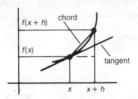

This may be viewed as finding the gradient of the tangent to the graph of f at the point x, by taking the limit of the gradients of a sequence of chords over increasingly small intervals.

For many families of functions, derivatives are simply deducible from the form of the original function. For example:

For $f(x) = Ax^n$ $f'(x) = nAx^{n-1}$.
So, if $f(x) = 3x^3 + 2x^2 + 1$ then $f'(x) = 9x^2 + 4x$.

Digit A single numeral (0, 1, 2, . . . 9) when used as part of the representation of a number in a place value system of writing numbers is called a digit. For example, 35780 – this number has five digits, the first being 3, and the last 0.

Digital A digital computer is one that represents numbers by a digital code (usually of ones and zeros), rather than by a continuously measurable quantity as do analogue devices.

Directed numbers Signed numbers (positive and negative) in which positive numbers are represented along a line relative to an origin 0, and negative numbers are similarly represented in the reverse direction.

$$\ldots \quad \frac{-3 \quad -2 \quad -1 \quad 0 \quad +1 \quad +2 \quad +3 \quad +4}{} \quad \ldots$$

In the directed number model, -3 can be thought of in two ways either as a position on the number line, or as a vector shift of 'three units left' along the line. Directed numbers can be used as a model for operations with integers. For example, $+1 + -3 = -2$ might be viewed as 'one step right then three steps left gives two steps left'.

Directrix A line that defines the shape of a curve of the **conic** family. For any point on a given curve the ratio of its distance from a fixed point (focus) to the directrix line is constant.

Discrete Quantities that can be measured by a counting process are called discrete. Discrete variables change in a discontinuous way.

The number of trees in a plantation is a discrete variable over a given period of time.

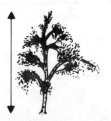

The height of a particular tree in the plantation is not a discrete variable – it is **continuous**.

Discriminant For a **quadratic equation**
$x^2 + bx + c = 0$ the quantity $b^2 - 4ac$ is called the discriminant.

If: $b^2 - 4ac > 0$ the equation has two real roots
$b^2 - 4ac = 0$ the equation has coincident real roots
$b^2 - 4ac < 0$ the equation has **imaginary** roots

Displacement A displacement is a **vector** quantity representing a change of position.

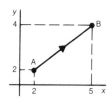

The displacement from A(2,2) to B(5,4) can be represented by the vector $\begin{pmatrix} 3 \\ 2 \end{pmatrix}$ and is of magnitude √61 at 33.7° to the x-axis.

Distribution The arrangement and frequencies of a set of values of a statistical variable is called a distribution. Certain distributions are more common than others, for example, Normal, Poisson, Binomial, etc.

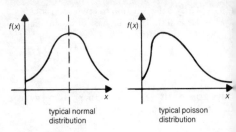

typical normal distribution

typical poisson distribution

Distributive law When two binary operations $\oplus$ and $*$ are defined on a set S, and for any three elements x, y, z in S

$$x*(y \oplus z) = (x * y) \oplus (x * z)$$

$*$ is said to distribute over $\oplus$.

In the arithmetic of real numbers multiplication distributes over addition. For example:

$$7 \times (4 + 3) = (7 \times 4) + (7 \times 3)$$

Addition does not, however, distribute over multiplication. For example:

$$5 + (3 \times 6) \neq (5 + 3) \times (5 + 6)$$

Dividend A number that is to be divided by another is called a dividend. For example, in 2468 ÷ 3, 2468 is the dividend.

Division The inverse operation to multiplication. One of the basic operations of arithmetic. The **quotient** of two numbers is determined by this operation which is associated with the process of sharing one quantity into a number of equal parts.

Divisor In a division problem the quantity by which the **dividend** is to be divided is called the divisor. The **factors** of a whole number are sometimes called divisors. For example, in 2468 ÷ 3, 3 is the divisor.

Dodecagon A **polygon** having twelve sides, hence twelve interior angles, is called a dodecagon.

Dodecahedron A **polyhedron** having twelve faces is called a dodecahedron.

A regular dodecahedron has for its faces twelve regular **pentagons**.

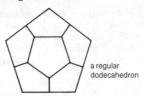

a regular
dodecahedron

Domain The set of objects which act as inputs to a **function** is called the domain of the function. For example, the function of $y = \sqrt{x}$, the domain is normally taken to be the set of possible values of the variable x, i.e., non-negative real numbers.

Dual The dual of a network is another network i

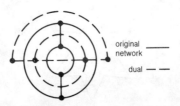

original
network

dual ─ ─ ─

which the regions and arcs of the former correspond to the nodes and arcs of the latter.

Networks and their duals share various common properties.

More generally in plane and solid geometry figures that can be obtained from one another by interchanging particular elements (For example, points ↔ lines) are called dual figures.

Theorems proved about one figure provide analogous theorems about the dual.

Duodecimal   The number system employing the **base** of twelve is called the duodecimal system.

Two extra numerals more than the familiar 0, 1, 2, . . . 9 are needed to represent ten and eleven in this system.

Base ten		Duodecimal
41	↔	35
tens units		twelves units

e   The symbol given to the irrational number which is the base of the natural logarithm function. A fundamental constant of both mathematics and the physical world.

The size of any quantity that grows or decays at a rate proportional to its size at a given time is related to the number e.

e = limit of $(1 + 1/n)^n$ as n increases. To ten places of decimals $e \simeq 2.7282818284$.

A unique property of the function $y = e^x$ is that the gradient of its graph for any value of x is identical to the value of the function for that value.

Eccentricity For a **conic**, the ratio e of the distance of any point on the curve from the **focus**, to the distance of the same point from the **directrix**, is called the eccentricity.

For $e < 1$ the conic is an **ellipse**

$e < 1$

For $e = 1$ the conic is a **parabola**

$e = 1$

$e > 1$

For $e > 1$ the conic is a **hyperbola**

Edge A line forming the intersection of two faces of a **polyhedron** or other solid is called an edge.

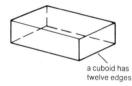

a cuboid has
twelve edges

a tetrahedron
has six edges

71

Element The individual members of a **set** are called its elements. For example:

Red is an element of the set of primary colours.
7 is an element of the set of odd numbers.
The symbol **ε** means 'is an element of' in set notation, hence 7 ε {odd numbers}.

Elevation The angle of elevation of an object from an observer viewing the object from below, is the angle between the line joining the object and the observer, and the horizontal plane.

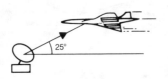

The angle of elevation of the plane from the radar station is 25°.
   In geometrical and mechanical drawings an elevation is a view of a solid object often from the front or side of the object.

72

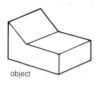

object

front
elevation

Ellipse A plane curve belonging to the **conic** family. An ellipse has an **eccentricity** e of less than 1.

If a cone is cut by a plane to form a cross section which is a closed curve, the cross section is an ellipse.

An ellipse can also be thought of as the locus of a point in a plane that moves such that the sum of the distances of the point from two fixed points, called the foci, is constant.

An ellipse has two axes of symmetry called the major and minor axes.

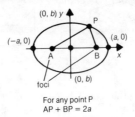

For any point P
AP + BP = 2a

For an ellipse with major axis extending from $(-a, 0)$ to $(a, 0)$, and minor axis from $(0, -b)$ to $(0, b)$ the cartesian equation is:

$$\frac{x^2}{a^2} + \frac{y^2}{b^2} = 1$$

A circle is a particular case of an ellipse where $e = 0$.
The area enclosed by an ellipse is equal to πab.

Empty set The empty set, which is denoted by the symbol Φ, is a set which contains no members. For example, the set of female kings of England $= \Phi$.

74

Enlargement An enlargement is a transformation of plane or solid shapes in which shapes are mapped onto **similar** shapes using a centre of enlargement.

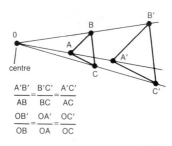

$$\frac{A'B'}{AB} = \frac{B'C'}{BC} = \frac{A'C'}{AC}$$

$$\frac{OB'}{OB} = \frac{OA'}{OA} = \frac{OC'}{OC}$$

The **scale factor** of an enlargement is the ratio of corresponding lengths on the object to those on the image. Alternatively the scale factor can be viewed as the ratio of distances between the centre and the object and corresponding distances between the centre and the image.

Enlargements centred at the origin are linear transformations and can be represented in matrix form. For example: in a plane

$$\begin{pmatrix} 3 & 0 \\ 0 & 3 \end{pmatrix} \quad \text{or} \quad \begin{pmatrix} -2 & 0 \\ 0 & -2 \end{pmatrix}$$

Envelope A plane curve which is tangential to each member of a whole family of curves is called an envelope to that family of curves.

Here the family of curves is of circles of radius a whose centres are all distance b from a fixed point 0.

This family has two envelopes, themselves circles, one of radius $(b + a)$ and the other of radius $(b - a)$.

Epicycloid An epicycloid is the path traced by a point P on the circumference of a circle which itself rolls on the circumference of another circle.

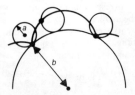

An epicycloid forms 'arches' around the fixed circle. If the radii of the rolling and fixed circles are a and b respectively, the curve has n 'arches' when $b = na$.

Equal Quantities which are alike in certain respects are said to be equal in those respects. The symbol $=$ is used to denote equality.

The relationship of equality is an example of an **equivalence** relation.

Equation An equation is a statement of an equality relationship between two quantities or expressions. A large part of mathematics is concerned with the study of methods of solution for various types of equations. For example:

$$(x + 1)^2 = x^2 + 2x + 1$$

is an equation expressing an identity relationship which holds for any numerical value of x.

$$3x + 1 = 10$$

is a simple **linear** equation which is conditional upon the value of x.

The solution of this equation is $x = 3$.

Other examples of common types of simple equations include **quadratic** and **simultaneous** equations.

Equator A line of **latitude** which is a great circle.

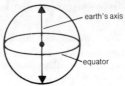

The equator is the circle produced on the earth' surface by taking a section through the centre of th sphere, perpendicular to the earth's axis.

The equator is the reference line for positions o latitude on the earth's surface. For example, latitud 60° north means 60° north of the equator.

Equilateral An equilateral **polygon** is one wit all its sides equal in lengths.

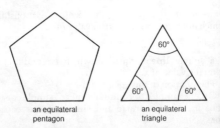

an equilateral   an equilateral
pentagon   triangle

An equilateral triangle also necessarily has equal interior angles. This is however not generally true of all equilateral polygons.

Equilibrium When the system of forces acting on a body are balanced and have a zero resultant, the body is said to be in a state of equilibrium.

This marble, balanced on an upturned cup, is in a state of unstable equilibrium. When displaced the marble will move away from the position of equilibrium.

This marble resting in the cup is in a state of stable equilibrium. When displaced the marble will move back to the position of equilibrium.

Equivalence An equivalence relation R between the members of a set has three properties:

(i) It is reflective: cRc for elements c in the set.
(ii) It is symmetric: If aRb then bRa.
(iii) It is transitive: If dRe and eRf then dRf.

For example, the relation 'is similar to' defined on a set of plane figures is an equivalence relation.

An equivalence relation splits up a set into distinct subsets called equivalence classes.

Equivalent Two fractions are called equivalent if they represent the same rational number, that is they can be **cancelled** to produce a common fraction

The fractions a/b and c/d are equivalent when $ad = bc$.

For example, $\frac{3}{4}$, $\frac{6}{8}$, $\frac{9}{12}$, $\frac{21}{28}$ and $\frac{315}{420}$ are all equivalent.

Euclidean After the famous Greek mathematician Euclid. A term often used to describe the ordinary space of three dimensions.

Euler's formula The characteristic equation linking certain properties of plane networks and **polyhedra**.

or networks

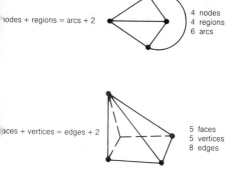

nodes + regions = arcs + 2

4 nodes
4 regions
6 arcs

aces + vertices = edges + 2

5 faces
5 vertices
8 edges

ven Numbers that have 2 as a factor are called
ven. Even numbers are the multiples of 2: 2, 4, 6, 8
. .

xpansion The extended or expanded form of
a expression. For example, the expansion of $(x + 1)^3$
$x^3 + 3x^2 + 3x + 1$.

xponent An exponent is the numerical symbol
ed to indicate the raising of a **base** number to some
dicated **power**. For example, the exponent in 3^4 is

Exponential A term applied to functions of the type $f(x) = a^x$. All functions of this type have certain features in common. For example, they are positive for all x. They tend to 0 as x becomes increasingly negative. Their graphs cut the y axis at 1 that is $f(0) = 1$.

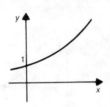

The function $f(x) = \mathbf{e}^x$, in particular, is known as the exponential function. These functions are associated with situations of natural growth and decay. Their inverses are **logarithmic** functions.

Extrapolate The use of past figures to try and predict the future, is called extrapolation. In the graph below we try to predict the population of Great Britain in 1911 from the figures over the period 1821–91.

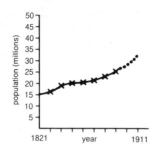

face The flat surfaces of a **polyhedron** are called its faces.

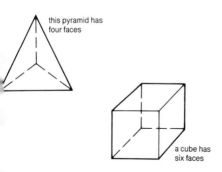

this pyramid has four faces

a cube has six faces

83

Factor The factors of a number are all the numbers which divide into it exactly. For example:

> factors of 12 are 1, 2, 3, 4, 6, 12
> factors of 17 are 1, 17
> factors of 32 are 1, 2, 4, 8, 16, 32

The factors of a **polynomial** are other polynomials that divide into it exactly. For example:

> factors of $x^3 - 8$ are $x - 2$ and $x^2 + 2x + 4$
> factors of $x^3 + 1$ are $x + 1$ and $x^2 - x + 1$

Factor theorem The factor theorem states that if $P(x)$ is a **polynomial** and $P(a) = 0$, then $x - a$ is factor of the polynomial. For example:

if $P(x) = x^3 - 6x^2 + 11x - 6$
$P(1) = 1 - 6 + 11 - 6 = 0 \Rightarrow x - 1$ is a factor
$P(2) = 8 - 24 + 22 - 6 = 0 \Rightarrow x - 2$ is a factor
$P(3) = 27 - 54 + 33 - 6 = 0 \Rightarrow x - 3$ is a factor

so the factors of $x^3 - 6x^2 + 11x - 6$ are $(x - 1)$, $(x - 2)$ and $(x - 3)$ and
$x^3 - 6x^2 + 11x - 6 \equiv (x - 1)(x - 2)(x - 3)$.

Factorial Factorial n, written as $n!$ is defined the product of all the integers 1, 2, 3, . . . up to and including n. For example:

$$5! = 5 \times 4 \times 3 \times 2 \times 1 = 120$$
$$4! = 4 \times 3 \times 2 \times 1 \quad = 24$$
$$3! = 3 \times 2 \times 1 \quad\quad = 6$$
$$2! = 2 \times 1 \quad\quad\quad = 2$$
$$1! = 1 \quad\quad\quad\quad = 1$$

Using the above pattern, we may deduce a value of 1 for $0! = 1$.

Fahrenheit The fahrenheit scale is a scale of measurement of temperatures. In this system, the boiling and freezing points of water are 212° and 32° respectively. The normal body temperature of a person is 98.4° fahrenheit.

A temperature in fahrenheit may be changed to a temperature in **centigrade**, the other common scale of measurement, using the formula:
$C = 5/9 \ (F - 32)$.

Farey The farey sequence of fractions of order 4 is the set of all simplified fractions less than 1 whose **denominators** are 4 or less, arranged in order of size that is, ¼, ⅓, ½, ⅔, ¾.

For example the farey sequence of order 7 is:

⅐, ⅙, ⅕, ¼, 2/7, ⅓, ⅖, 3/7, ½, 4/7, ⅗, ⅔, 5/7, ¾, ⅘, ⅚, 6/7

Fibonacci The fibonacci sequence of numbers is the sequence 1, 1, 2, 3, 5, 8, 13, 21, 34, etc. Each member of the sequence is the sum of the two preceding numbers.

Foot A unit of length. One foot is equal to twelve inches or just over thirty centimetres.

Formula A general rule or result stated in algebraic equation form.
Examples:

$P = 2(l + w)$ is a formula for the perimeter of a rectangle, l and w stand for the length and width of the rectangle.

$F = 9/5\ C + 32$ is a formula linking temperatures in degrees fahrenheit (F) and centigrade (C).

$A = \pi R^2$ is a formula for the area of a circle in terms of its radius.

In the above examples the letters P, F and A are said to be the subjects of the formulae. We may change the subject by rearranging the formula. For example:

$$P = 2(l + w)$$
$$\frac{P}{2} = l + w$$
$$\frac{P}{2} - l = w$$

$$A = \pi R^2$$
$$\frac{A}{\pi} = R^2$$
$$\sqrt{\frac{A}{\pi}} = R$$

w is now the subject R is now the subject

86

Fraction A fraction is part of a whole. The fraction 'three quarters' is denoted by the **rational number** ¾. The bottom number is often called the **denominator** and the top number is called the **numerator**.

The shaded part represents ⅝ of the whole circle, since the circle has been divided into eight equal parts, five of which are shaded.

addition:

$$4⅝ + 3⁷\!/_{12} = 7\frac{15 + 14}{24} = 7^{29}\!/_{24} = 8^5\!/_{24}$$

subtraction:

$$4⅜ - 1⁷\!/_{12} = 3\frac{\overset{24}{9} - 14}{\underset{2}{24}} = 2^{19}\!/_{24}$$

multiplication:

$$3\tfrac{3}{4} \times 1\tfrac{1}{9} = \frac{\overset{5}{15}}{\underset{2}{4}} \times \frac{\overset{5}{10}}{\underset{3}{9}} = \frac{25}{6} = 4\tfrac{1}{6}$$

division:

$$3\tfrac{3}{4} \div 1\tfrac{1}{9} = \frac{15}{4} \div \frac{10}{9} = \frac{\overset{3}{15}}{4} \times \frac{9}{\underset{2}{10}} = \frac{27}{8} = 3\tfrac{3}{8}$$

A fraction may be changed into a decimal by dividing the numerator by the denominator. For example:

$$\tfrac{3}{8} = 0.375 \qquad 8\overline{)3.0^{6}0^{4}0}\;\;{}^{0.375}$$

Frequency The frequency of an event is how many times it has occurred. For example, a die is thrown ten times and the results are 1, 2, 5, 6, 5, 4, 3, 1, 2, 1. The frequency of 1 is three since three of the throws were 1s. The results may be written in a frequency table as below:

Score	1	2	3	4	5	6
Number of throws (frequency)	3	2	1	1	2	1

Frustum The frustrum of a solid shape is any part of the solid contained between two parallel planes that cut the solid.

frustum
of a cone

frustum of a
pyramid

Function A function is a relation between two sets called the **domain** and **range** in which each member of the domain is related to precisely one member of the range, called its **image**.

Functions are mathematical abstractions of systems of two or more **variables** which are related to one another. For example:

$x \rightarrow 2x + 3$ Describes a function from the set of real numbers to itself.

$x \rightarrow$ 'Number of factors of x' describes a function from the set of positive integers to itself.

$x \rightarrow 1/x$ Describes a function from the set of non-zero real numbers (0 would have no image) to itself.

Geometric mean The geometric mean of a set of n numbers is the nth **root** of the **product** of all of the numbers. For example:

geometric mean of 2, 4, 8 is $\sqrt[3]{2 \times 4 \times 8} = \sqrt[3]{64} = 4$

geometric mean of 2, 3, 6, 12, 18, is

$\sqrt[5]{2 \times 3 \times 6 \times 12 \times 18} = 6$

Geometric progression A geometric progression is a **sequence** of numbers, in which each term is a constant **multiple** of the preceding term. This multiple is called the common ratio. For example:

2, 6, 18, 54, 162 common ratio is 3
64, 32, 16, 8, 4, 2 common ratio is ½
¼, −1, 4, −16, 64 common ratio is −4

The sum of the terms of the general geometric progression $a + ar + ar^2 + ar^3 \ldots + ar^{n-1}$ is given by the formula:

$$S_n = \frac{a(r^n - 1)}{r - 1} \quad \text{or} \quad \frac{a(1 - r^n)}{1 - r}$$

(a is the first term of the series and r is the common ratio)

The sum of an infinite geometric progression can be calculated if the common ratio is between −1 and +1.

$$S_\infty = \frac{a}{1 - r}$$

Geometry Geometry is a branch of mathematics which involves the study of special properties of figures and solids.

Gradient The gradient of a line is a measure of its steepness.

The gradient is $\frac{4}{2} = 2$. Lines sloping this way have a positive gradient.

The gradient is $\frac{-5}{2} = -2\frac{1}{2}$. Lines sloping this way have a negative gradient.

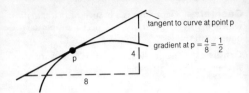

The gradient of a curve at a particular point is the gradient of the **tangent** drawn to the curve at that point.

Gram A unit of mass. There are one thousand grams in a kilogram, and just over twenty-eight grams in one ounce.

Graph A graph is a diagramatic representation of the relationship between various quantities.

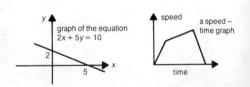

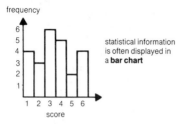

frequency

statistical information
is often displayed in
a **bar chart**

score

Often, **coordinates** and **axes** are used.

Great circle   A great circle is a circle drawn on
the earth's surface that has the same **radius** as that of
the earth. For example, the **equator** is a great circle,
all longitude lines are parts of great circles.

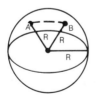

Given two points on the surface of the earth, the
shortest distance between them (along the surface) is
along a great circle route. (See diagram above.)

Group A set of numbers, letters, etc., is said to form a group under an operation ∗ if the four following properties hold.

1) The set is **closed** under the operation ∗.
2) The operation ∗ is **associative**.
3) An **identity** element exists and is unique.
4) Each element of the set has a unique **inverse**.

For example, all 2 by 2 **matrices** with non-zero **determinant** from a group under matrix multiplication.

The numbers 1, 5, 7, 11 form a group under multiplication **modulo** 12.

Helix A helix is a curve which lies on the surface of a cylinder or cone and cuts the shape at a constant angle.

Hemisphere A hemisphere is half of a **sphere** and is formed when a sphere is cut into two, the cut passing through the sphere's centre.

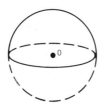

Heptagon A heptagon is a seven sided **polygon**. The sum of all the interior angles of a heptagon is 900°.

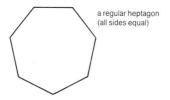

a regular heptagon (all sides equal)

Hero's formula A formula for the area of a triangle.

$$A = \sqrt{s(s - a)(s - b)(s - c)}$$

$s = \frac{1}{2}(a + b + c)$

Hexagon A hexagon is a **polygon** having six sides.

The diagram shows a regular hexagon, each of whose interior angles is equal to 120°.

Highest common factor The highest common factor (or HCF) of two or more numbers is th hem. For example:

> HCF of 6, 8, 12 is 2
> HCF of 10, 15, 20 is 5
> HCF of 20 and 45 is 5
> HCF of 42 and 56 is 14

Histogram A histogram is similar to a **bar chart** except that the **frequency** of the bar is represented by its area rather than its height.

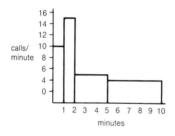

length of phone call (minutes)	Number of calls	calls/minute
0–1	10	$10 \div 1 = 10$
1–2	15	$15 \div 1 = 15$
2–5	15	$15 \div 3 = 5$
5–10	20	$20 \div 5 = 4$

97

Homogeneous A homogeneous **polynomial** is one whose terms are all of the same **degree**.

$x^3 + x^2y + 3xy^2 + 7y^3$ is homogeneous of degree 3

or

$2x^2 + 3xy - y^2$ is homogeneous of degree 2.

Note: the degree of the term is obtained by adding together the powers of x and y, so x^3y^4 is of degree 7.

Similarly a homogeneous **equation** may be defined. For example:

$x^3 + x^2y + 3xy^2 + 7y^3 = 0$ is homogeneous of degree 3.

Hooke's law Hooke's law states that the extension of a spring is proportional to the load put on it (provided that the spring returns to its unstretched length when the load is removed).

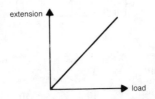

$E \propto L$ or $E = k \times L$ where k is a constant depending on the properties of the material of the spring.

Horizontal   A line is said to be horizontal if it is parallel to the earth's skyline.

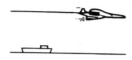

Hour   A unit of time. One hour = sixty minutes, and there are twenty-four hours in a day.

Hyperbola   A hyperbola is the **locus** of a point which moves so that the ratio of its distance from a fixed point (called the focus) to its distance from a fixed line (called the **directrix**) is greater than 1.

The ratio:

$$\frac{PF_1}{PO} > 1$$

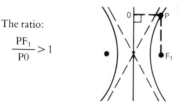

called the **eccentricity** of the hyperbola.

A hyperbola has two foci (plural of focus). The two dotted lines in the diagram above are called the **asymptotes** of the hyperbola.

Hypocycloid A hypocycloid is the **locus** of a point P fixed on the **circumference** of a circle which rolls on the inside of a given fixed circle.

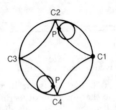

C_1, C_2, C_3, C_4 are called **cusps**. In the diagram the fixed circle has a diameter four times that of the rolling circle.

Hypotenuse Hypotenuse is the mathematical term for the longest side of a right-angled triangle. The hypotenuse is always opposite the right angle.

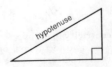

Hypothesis A hypothesis is an assumption which is made on the basis of past observations, which is then used as a condition when proving the truth of some further proposition.

Icosahedron An icosahedron is a **polyhedron** having twenty faces. A regular icosahedron is made up of twenty equilateral triangles.

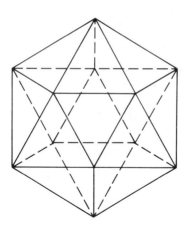

Identity An identity element is a member i of a set s which, when combined with any other member of the set, leaves it unchanged:
$i * x = x * i = x$ for $x \in s$

For example, for addition, 0 is the identity since $0 + 7 = 7$, $0 + -2 = -2$, $0 + 150 = 150$, etc.

For multiplication, 1 is the identity since $1 \times 7 = 7$, $1 \times -2 = -2$, $1 \times 19 = 19$, etc.

For **sets** under the operation of **union**, Φ, (the empty set) is the identity:
$\Phi \cup \{a, b, c, d\} = \{a, b, c, d\}$. $\Phi \cup A = A$ (for any A).

For 2 × 2 **matrices** the identity for multiplication is:

$$\begin{pmatrix} 1 & 0 \\ 0 & 1 \end{pmatrix} \text{ since } \begin{pmatrix} 1 & 0 \\ 0 & 1 \end{pmatrix} \begin{pmatrix} a & b \\ c & d \end{pmatrix} = \begin{pmatrix} a & b \\ c & d \end{pmatrix}$$

In the operation table below, b is the identity.

	a	b	c	d
a	d	a	b	c
b	a	b	c	d
c	b	c	d	a
d	c	d	a	b

Note: the 'b' row is the same as the top row of the table, and the 'b' column is the same as the left column of the table.

Image The result of applying a **function** to a particular member of the **domain** is called the image of that element. For example:

$x \rightarrow 2x + 3$, $2 \rightarrow 7$. The number 7 is said to be the image of the number 2 which is called the object.

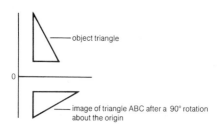

object triangle

image of triangle ABC after a 90° rotation about the origin

In geometry the term is often applied to the result of some transformation.

Imaginary The imaginary part of a complex number is the non-real part of the number. For example, the imaginary part of $3 + 4i$ is 4.

Improper An improper fraction is one whose **numerator** is greater than its **denominator**. For example, $^{17}\!/_6$, $^8\!/_3$, $^{13}\!/_4$, etc.

An improper fraction may be changed into a mixed number. For example, $^{17}\!/_6 = 2^5\!/_6$, $^{13}\!/_4 = 3^1\!/_4$.

Inch A unit of length. There are twelve inches in one foot and one inch is just bigger than 2½ cm.

Incircle In the triangle below, the angles have been **bisected**. All these bisectors meet at the same point, 0, called the incentre of the triangle. With 0 as centre it is possible to draw a circle that just touches each of the sides of the triangle. This circle is called the incircle, or inscribed circle of the triangle.

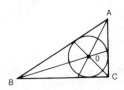

Independent The equation $y = 2x + 5$ defines y as a function of x. x is said to be the independent variable of the function (and y is the dependent variable).

In the equation $S = 2t^2 + 3t - 5$, the independent variable is t.

Index (plural indices) In the number 2^3, 3 is called the index or power.

When numbers are multiplied or divided, the rules given below apply:

$x^a \times x^b = x^{a+b}$ (add indices when multiplying)
$x^a \div x^b = x^{a-b}$ (subtract indices when dividing)
$(x^a)^b = x^{ab}$

Using these rules, negative and fractional indices may be deduced:

$$x^{-a} = \frac{1}{x^a}, \; x^{1/n} = \sqrt[n]{x}, \text{ and } x^{a/b} = \sqrt[b]{x^a} \text{ or } (\sqrt[b]{x})^a$$

also $x^\circ = 1$ for any number x

$$2^{-5} = \frac{1}{2^5} = \frac{1}{32}, \; 125^{1/3} = \sqrt[3]{125} = 5,$$
$$8^{2/3} = \sqrt[3]{8^2} = \sqrt[3]{64} = 4$$

When a number (for example, 16) is written as $16 = 2^4$, it is said to be written in index form.

Similarly
$$360 = 2 \times 2 \times 2 \times 3 \times 3 \times 5$$
$$= 2^3 \times 3^2 \times 5^1.$$

Inequality An inequality or ordering is a mathematical statement that one quantity is greater or less than another. For example:

$3 > 2$ (3 is greater than 2)
$1 < 4$ (1 is less than 4)
$x \geq 1$ (x is greater than or equal to 1)
$y \leq 2$ (y is less than or equal to 2)

Inequalities remain true if the same quantity is added to or subtracted from both sides, or if both sides are multiplied or divided by the same positive number. However multiplying or dividing by a negative number changes the sense of the inequality:

5 > 2	adding 3	8 > 5	
	subtract 4	1 > −2	all still true
	× by 3	15 > 6	
	÷ by 2	2½ > 1	

but

	× by −4	−20 < −8	direction of sign
	÷ by −2	−2½ < −1	must change to make the statement true

Infinite   A quantity is said to be infinite if it is larger than any fixed limit. For example, the set, {1, 2, 3, 4, . . .} has an infinite number of members. But the set, {a, b, c, . . . y, z} has a finite number of members, since they can be counted (twenty-six members).

The symbol ∞ is used to represent an infinite value, or infinity.

Inflection   A point of inflection on a curve is where the curve changes from being **concave** upwards to concave downwards (or vice versa). The point P below is said to be a **stationary point** of inflection since the **gradient** at P is zero. Q is a non-stationary inflection since the gradient is non-zero at Q.

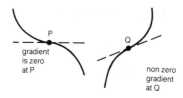

gradient is zero at P

non zero gradient at Q

Integer An integer is a whole number. For example:

$$\dots -5, -4, -3, -2, -1, 0, +1, +2, +3, +4, +5, \dots$$

negative integers   positive integers

Integral The integral of the function $f(x)$ is defined as the limit as δx approaches zero, of the sum of the areas of the rectangles under the curve (see diagram below).

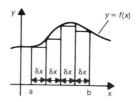

The area under the curve between the values $x = a$ and $x = b$ is written as:

$$\int_a^b f(x)\, dx$$

This is called a definite integral and a and b are known as the lower and upper limits of the integral.

Any function whose **derivative** is $f(x)$ is called and indefinite integral of $f(x)$. For example, $3x^2 + 7$, $3x^2 - 14$, $3x^2 + K$ are all indefinite integrals of $6x$.

Integration   Integration is the process of finding a definite or indefinite integral. Integration may be thought of as the inverse of **differentiation**.

Some important results are:

$$\int x^n dx = \frac{1}{n + 1} x^{n + 1} + K$$

$$\int \sin x \, dx = - \cos x + K \text{ and}$$

$$\int \cos x \, dx = \sin x + K$$

Intercept   The intercept of a straight line on an axis is the distance from the origin to the point where the line cuts the axis.

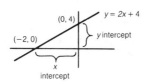

In the diagram above, for example, the x and y intercepts are 2 and 4 respectively.

Interest   Interest is a charge made for borrowing money. There are two main ways of calculating interest.

Simple interest is where the interest is worked out over the whole period, just on the capital borrowed:

$$\text{Simple interest} = \frac{P \times R \times T}{100}$$

P = amount borrowed
R = rate of interest (per cent)
T = time of loan

When the interest due is added onto the amount borrowed, and then itself earns interest, it is called **compound interest**:

Interest + capital at compound interest of six per cent means that $P \times 1.06^n$ is owed after n years.

Interpolation Interpolation is the process of finding the value of a function at a point using two known values on either side of the point.

For example, using linear interpolation the predicted value of $f(c)$ using the known values $f(a)$ and $f(b)$, is as shown on the diagram below.

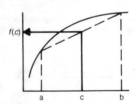

Interquartile range The interquartile range of a distribution of numbers is the difference between the upper and lower **quartiles** of the distribution.

I.Q.R. = (upper quartile) − (lower quartile)

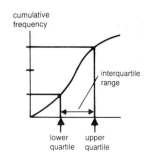

Intersection The intersection of two lines or curves is the set of points in common to the two.

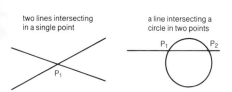

two lines intersecting in a single point

a line intersecting a circle in two points

The intersection of two sets is the set of elements that they have in common.

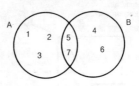

A = {1, 2, 3, 5, 7} ⇒ A ∩ B = {5, 7}
B = {4, 5, 6, 7}
Note: the symbol ∩ means intersection.

Interval If x is a real number and can take all values between -2 and $+3$ inclusive, we say that x lies in the interval -2 to $+3$. This may be represented in a diagram using a number line.

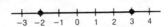

The diagram above shows the 'end points' are included. Algebraically this interval is written as:

$$\{x : -2 \leqslant x \leqslant 3\} \text{ or } [-2, 3]$$

and is sometimes called a closed interval, since both end points are included.

An interval where the end points are not included is called an open interval:

The open interval: $\{x : -1 < x < 5\}$ or $]-1, 5[$

The diagram above shows the 'end points' are not included.

An interval may be half open and half closed:

$\{x : -2 \leqslant x < 4\}$ or $[-2, 4[$

The diagram above shows -2 included in the interval and 4 excluded.

Invariant A quantity or property is invariant under a transformation if it remains unchanged by the transformation. For example, the areas of plane figures remain invariant under certain transformations such as rotations, reflections, shears, translations.

113

Inverse If a set with an operation $*$ defined on it has an **identity** element e, then any element x for which there exists another element x^{-1} with $x * x^{-1} = e$ is said to have an inverse element. For example:

The inverse of $+7$ under addition is -7.

The inverse of 8 under multiplication is $\frac{1}{8}$.

The inverse of the matrix $\begin{pmatrix} 5 & 3 \\ 6 & 4 \end{pmatrix}$ under

multiplication is $\begin{pmatrix} 2 & -1\frac{1}{2} \\ -3 & 2\frac{1}{2} \end{pmatrix}$.

The inverse of the function $f(x) = \dfrac{x-1}{2}$ is
$$f^{-1}(x) = 2x + 1.$$

The inverse of the plane transformation 'rotat through $+90°$ about $(0, 0)$' is 'rotate through -90 about $(0, 0)$'.

Involute The involute of a circle is the curv described by the end of a thread as it is unwound from a fixed spool (keeping the thread taut all th time).

The distances $d_1, d_2, d_3, \ldots$ are equal to $\frac{1}{4}, \frac{1}{2}, \frac{3}{4}$ etc. of the circumference of the circle.

114

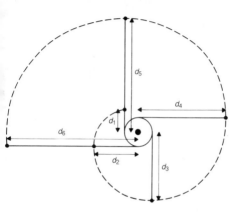

rrational An irrational number is one that is not **ational**, that is, it cannot be expressed as a fraction, r written in precise form as a decimal. For example, r, e, $\sqrt{2}$, $\sqrt{3}$, $1 - \sqrt{7}$. All the irrational umbers together with all the rational numbers nake the set of real numbers.

It can be shown that $\sqrt{2}$ is an irrational number s follows:

Suppose $\sqrt{2}$ is rational

i.e. $\sqrt{2} = \dfrac{a}{b}$ (where this fraction has been fully cancelled)

$\therefore \quad 2 = \dfrac{a^2}{b^2}$

$\therefore \quad 2b^2 = a^2$ since $2b^2$ is always even, then a must be even, e.g., $a = 2p$

$\therefore \quad 2b^2 = (2p)^2$ (replacing a by $2p$)

$\therefore \quad 2b^2 = 4p^2$

$\therefore \quad b^2 = 2p^2$ since $2p^2$ is always even, then b must be even

We have proved both a and b are even, which means that they have a common factor of 2, so the fraction a/b cancels by 2. This is clearly a contradiction, therefore the original assumption that $\sqrt{2}$ could be expressed as a fraction must have been invalid.

Isometric Isometric graph paper is paper ruled into equilateral triangles rather than squares.

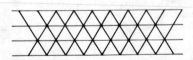

Isometry An isometry is a transformation under which all lengths are **invariant**. For example, trans

116

lations, rotations, reflections are isometries; shears, enlargements, stretches are not.

Isometries that preserve sense also are called direct isometries, and those that change sense are called opposite isometries.

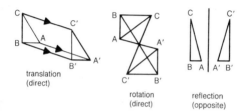

translation
(direct)

rotation
(direct)

reflection
(opposite)

Isomorphic   Two sets are said to be isomorphic if they have the same structure.

(A)

+	0	1	2	3
0	0	1	2	3
1	1	2	3	0
2	2	3	0	1
3	3	0	1	2

(addition)
modulo 4

(B)

	0°	90°	180°	270°
0°	0	90	180	270
90°	90	180	270	0
180°	180	270	0	90
270°	270	0	90	180

(combining)
rotations

Note: 0 in table (A) above, corresponds to 0° in table (B). 1 in table (A) corresponds to 90° in table (B), etc.

Also note that each set comprises **identity**, one self **inverse** element and an inverse pair.

(C)	a	b	c	d
a	a	b	c	d
b	b	a	d	c
c	c	d	a	b
d	d	c	b	a

This is not isomorphic to the other two systems above since every element is self inverse.

(D)	1	2	3	4
1	1	2	3	4
2	2	4	1	3
3	3	1	4	2
4	4	3	2	1

(multiplication)
modulo 5

This set is isomorphic to the first two, (A) and (B) above, but the table may need rearranging to make this more obvious.

(E)	1	2	4	3
1	1	2	4	3
2	2	4	3	1
4	4	3	1	2
3	3	1	2	4

Isomorphism When two sets are isomorphic, the mapping that relates the members of one set with their corresponding members in the second set is called an isomorphism. That is, an isomorphism ϕ is a one–one mapping such that $\phi(a * b) = \phi(a) . \phi(b)$ for all a and b. For example, an isomorphism between the sets {0, 1, 2, 3} and {0°, 90°, 180°, 270°} (whose operation tables are given under **isomorphic**).

0→0°   (identities correspond)
1→90°
2→180° (self inverse members correspond)
3→270°

Or equally well, the isomorphism could have been 0→0°, 1→270°, 2→180°, 3→90°.

An isomorphism preserves structure, that is, if two members of the first set are combined, and their corresponding members in the second set are also combined, the two answers will correspond.

The set of positive real numbers under the operation of multiplication is isomorphic to the set of real numbers under addition using the isomorphism x→log x. This is the basis for the use of the logarithm function as a calculating aid. The study of isomorphic systems is an important branch of modern algebra, as results proved about one system can be applied to the second.

Isosceles An isosceles triangle is one that has two sides and two angles equal.

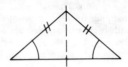

An isosceles trapezium is one whose two non-parallel sides are equal.

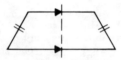

Each of the above figures has one line of symmetry (as shown in the diagrams).

Iterate To iterate means to repeat. So,

$$\frac{d^3 y}{dx^3}$$

is an iterated differential, since y is **differentiated** three times.

Iterative An iterative process involves the repetition of a sequence of operations to improve some result. For example:

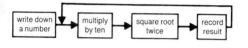

is an iterative way of finding $\sqrt[3]{10}$

Kilogram A unit of mass: 1 kilogram = 1000 grams (1 kg = 1000g). 1 kg is approximately 2.2 lbs.

Kilometer A unit of distance: 1 kilometre = 1000 metres (1 km = 1000 m). 1 km is approximately ⅝ of a mile.

Kite A kite is a **quadrilateral** having two pairs of adjacent sides equal.

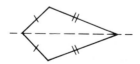

A kite has one line of symmetry. This diagonal

121

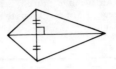

bisects the figure. The second diagonal is at right angles to the first diagonal but neither bisects the first nor the figure.

Knot 1) A knot is a unit of speed equivalent to one **nautical mile** per hour.

2) A topological knot is formed by looping and interweaving a piece of string and then tying the two ends together.

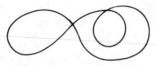

Königsberg bridge problem An historic problem in **topology** attributed to the famous Swiss mathematician Euler. It concerns the possibility of

122

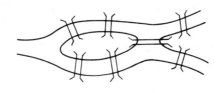

crossing each of the seven bridges in the town of Königsberg just once and returning to a starting point. By analysing the network of routes Euler showed the journey was not possible.

Latin square A Latin square is a square array of numbers or letters in which each number or letter appears once in every row and column. For example:

$$
\begin{array}{cccc}
1 & 2 & 3 & 4 \\
4 & 3 & 2 & 1 \\
3 & 1 & 4 & 2 \\
2 & 4 & 1 & 3 \\
\end{array}
\quad \text{but not} \quad
\begin{array}{cccc}
p & q & r & s \\
q & r & s & p \\
r & q & p & s \\
s & p & r & p \\
\end{array}
$$

The combination table for the members of a set which forms a group must be a Latin square.

Latitude A parallel of latitude is a circle drawn on the earth's surface, whose centre is on the line joining the north and south poles. All latitude lines are parallel to each other, hence the term, 'parallel' of latitude. The latitude of any point is given by the angle $\angle A0B$ where 0 is the centre of the earth and A and B are points on the required latitude line and the **equator** respectively. A and B are on the same **longitude** line. All latitude lines, apart from the equator, are **small circles**.

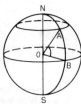

Limit A **sequence** is said to tend to a limit if its terms approach a definite value, getting ever closer as more and more terms are included. Similarly the sum of a **series** may tend to a limit. For example:

$1, 1\frac{1}{2}, 1\frac{3}{4}, 1\frac{7}{8}, 1\frac{15}{16}$ . . . limit is 2
$7.1, 7.01, 7.001, 7.0001$ . . . limit is 7
$4 + 2 + 1 + \frac{1}{2} + \frac{1}{4} + \frac{1}{8} + $ . . . limit of the sum is 8
$6 + 2 + \frac{2}{3} + \frac{2}{9} + \frac{2}{27} + $ . . . limit of the sum is 9
$a + ar + ar^2 + ar^3 + $ . . . limit of the sum is
$$a/1 - r\,(-1 < r < 1)$$

124

Line segment   A line segment is part of a straight line:

For example, AB is a line segment, being part of the line *l*. The points A and B do not necessarily have to be part of the line segment.

Linear   A general term pertaining to systems related in some way to straight lines.

A linear relationship between two variables *x* and *y*, is one that can be represented graphically by a straight line. Equations of such relationships (linear equations) can be written in the form $y = mx + c$.

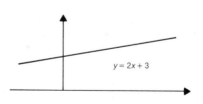

$y = 2x + 3$

The value of *m* gives the **gradient** of the line graph. The value of *c* gives the **intercept** of the graph on the *y* axis.

125

Litre A litre is a measure of volume equal to 1000 cubic centimetres. 1 litre = 1000 cm³. There are just over four and a half litres in one gallon.

Locus The path of a moving point.

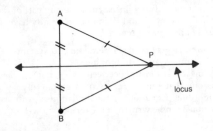

The locus of a point *p*, which moves so that it is equidistant from two fixed points A and B, is the perpendicular bisector of the line joining A and B.

Logarithm The logarithm of a number to a given **base**, is the power to which the base must be raised to give that number. For example:

$\log_{10} 100 = 2$ since $10^2 = 100$ (base = 10)
$\log_2 64 = 6$ since $2^6 = 64$ (base = 2)
$\log_5 0.008 = -3$ since $5^{-3} = 0.008$ (base = 5)

Logarithms to base 10 are called **common logarithms** and to base **e** (*e* = 2.718 . . .) are called **natural** or Naperian logarithms. The natural logarithm function, usually written $\log_e x$ or $\ln x$ is an important function in the study of **calculus**.

Before the introduction of calculators, logarithms were particularly useful since they reduced multiplication and division problems into problems of addition and subtraction respectively, hence saving time and tedious numerical working. For example:

To calculate 47.3 × 12.9

$\left. \begin{array}{l} \log_{10} 47.3 = 1.6749 \\ \log_{10} 12.9 = 1.1106 \end{array} \right\}$ adding logs we get 2.7855

Anti log 2.7855 = 610.2

Hence 47.3 × 12.9 ≃ 610.2

Longitude A longitude line is a semi-circle drawn on the surface of the earth joining the north and south poles. Longitude lines are parts of **great circles**, that is, circles having the same radius as that of the earth.

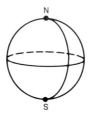

Just as **latitude** is measured in degrees north or south of the **equator**, so longitude is measured in degrees east or west of Greenwich, a fixed longitude line passing through Greenwich, London. (Known as the Greenwich meridian.)

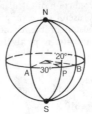

If in the diagram above, NPS represents the Greenwich line (0°) and angles P0A and P0B are 30° and 20° respectively, then all points on the line NAS are on the longitude 30°W line, and all points on line NBS are on the 20°E line.

Lower bound A lower bound of a set of numbers, is a number that is less than or equal to every member of the set. For example, {½, ⅓, ¼, ⅕ . . .} has a lower bound of 0.

Lowest 1) The lowest common multiple (LCM) of two or more numbers is the smallest number that they will all divide into exactly. For example:

LCM of 2, 3, 4 is 12
LCM of 18, 27 is 54
LCM of 8, 16 is 16
LCM of x, $2x$, x^2 is $2x^2$

2) The lowest common **denominator** of a set of fractions is the denominator that is the lowest common multiple of all the denominators. For example:

lowest common denominator of ¾, ⅚ and ⅓ is 12

lowest common denominator of $\dfrac{2}{x+1}$ and $\dfrac{4}{x^2-1}$ is x^2-1

3) A fraction is said to be in its lowest terms if it has been **cancelled** as much as possible. For example:

$\dfrac{3}{4}$, $\dfrac{5}{7}$, $\dfrac{1}{x-1}$, $\dfrac{3}{x^2}$ are in their lowest terms,

but

$\dfrac{6}{9}$, $\dfrac{20}{25}$, $\dfrac{x-1}{x^2-1}$, $\dfrac{3x}{x^4}$

are not since they will cancel down to

$\dfrac{2}{3}$, $\dfrac{4}{5}$, $\dfrac{1}{x+1}$ and $\dfrac{3}{x^3}$ respectively.

Magic square A magic square is a square array of numbers in which every row, column and the two diagonals add up to give the same total. For example:

129

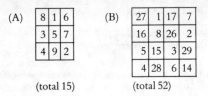

(A)		
8	1	6
3	5	7
4	9	2

(total 15)

(B)			
27	1	17	7
16	8	26	2
5	15	3	29
4	28	6	14

(total 52)

Example (B) above is rather special since in addition each block of four in the corners adds up to 52, the block of four in the centre adds up to 52 and the shorter diagonals (16 + 1 + 6 + 29) and (5 + 28 + 17 + 2) add up to 52. The four corners add up to 52 and the middle two numbers of the top and bottom rows add up to 52. Also the middle two numbers of the left and right columns add up to 52!

Magnitude The magnitude of a **vector** is the length of the vector, when represented in line segment form.

By Pythagoras $3^2 + 4^2 = 5^2$. So the magnitude is 5.

Major 1) The longer of the two axes of symmetry of an ellipse is called the major axis. The other is called the **minor** axis. For an ellipse with equation:

$$\frac{x^2}{a^2} + \frac{y^2}{b^2} = 1 \ (a > b)$$

The major axis has length $2a$.

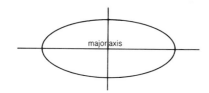

2) The shaded part of the diagram shows the major **sector** of the circle. The arc ACB is called the major **arc** of the circle.

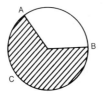

Mantissa When the **logarithm** of 1500 is written as 3.1761 the decimal part of the logarithm (i.e., .1761) is the mantissa. It is this part of the logarithm that determines the figures in the answer, but not the position of the decimal place. For example, a mantissa of .2858 means that the number will contain the figures 1931.

Mapping If the members of the set {8, 6, 4, 2} are each halved, a new set {4, 3, 2, 1} is formed. Halving transforms members of one set into members of another.

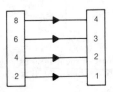

The diagram above illustrates what happens to the numbers 8, 6, 4, 2. We say that the members of one set are mapped onto the members of the other. The numbers 4, 3, 2, 1 are called the **images** of 8, 6, 4, 2.

A mapping is a special type of relation in which every member of the set has one image.

A mapping diagram is a diagram to illustrate a mapping.

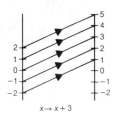

$x \rightarrow x + 3$

mapping diagram for the set $\{-2, -1, 0, 1, 2\}$
under the mapping $x \rightarrow x + 3$

Matrix A matrix is a rectangular array of numbers used to store information about particular mathematical systems. For example:

$$A = \begin{pmatrix} 3 & 5 \\ 2 & 4 \end{pmatrix} \qquad B = \begin{pmatrix} 6 & 3 & 1 \\ 5 & 2 & 7 \end{pmatrix}$$

$$C = \begin{pmatrix} 8 & 1 \\ 6 & 2 \\ 5 & 3 \end{pmatrix} \qquad D = \begin{pmatrix} 10 & 5 \\ 4 & 2 \end{pmatrix}$$

Matrix A has 2 **rows** and 2 **columns**. Matrix B has 2 rows and 3 columns.

Matrices of the same order (i.e., with the same number of rows and columns) can be added, by adding corresponding elements. For example:

133

$$A + D = \begin{pmatrix} 3 + 10, 5 + 5 \\ 2 + \ \ 4, 4 + 2 \end{pmatrix} = \begin{pmatrix} 13 & 10 \\ 6 & 6 \end{pmatrix}$$

Matrices A and B are compatible for multiplication because A has as many columns as B has rows. Multiplication involves the combining of the rows of A with the columns of B:

$$A \times B =$$
$$\begin{pmatrix} 3 \times 6 + 5 \times 5, 3 \times 3 + 5 \times 2, 3 \times 1 + 5 \times 7 \\ 2 \times 6 + 4 \times 5, 2 \times 3 + 4 \times 2, 2 \times 1 + 4 \times 7 \end{pmatrix}$$

$$= \begin{pmatrix} 43 & 19 & 38 \\ 32 & 14 & 30 \end{pmatrix}$$

Square matrices with non-zero **determinants** have multiplicative **inverses**:

The inverse of $\begin{pmatrix} a & b \\ c & d \end{pmatrix} = \dfrac{1}{ad - bc} \begin{pmatrix} d & -b \\ -c & a \end{pmatrix}$

For example: $A^{-1} = \begin{pmatrix} 2 & -2\frac{1}{2} \\ -1 & 1\frac{1}{2} \end{pmatrix}$

Matrices are a very powerful tool in the study o the solutions of **linear** equations.

Maximum

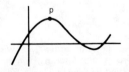

The graph shown above is said to have a maximum point at the point P.

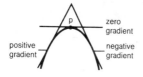

The **tangent** drawn to the curve at the point P would have zero **gradient**. Just to the left of P, the tangent would have a positive gradient and just to the right it would be negative.

Mean   The arithmetic mean of a set of numbers is the sum of all the numbers divided by the number of figures in the set. (Often called the '**average**'). For example, the mean of 2, 6, 8, 9, 12 is:

$$\frac{2 + 6 + 8 + 9 + 12}{5} = \frac{37}{5} = 7.4.$$

score on a dice	1	2	3	4	5	6	(total 20
number of throws	5	3	4	2	3	3	throws)

The mean score is:

$$\frac{(1 \times 5) + (2 \times 3) + (3 \times 4) + (4 \times 2) + (5 \times 3) + (6 \times 3)}{20}$$

$$= \frac{64}{20} = 3.2.$$

Mean deviation   The mean deviation (or strictly speaking – the mean absolute deviation from the arithmetic mean) of a list of numbers is defined as:

$$\frac{\text{sum of all deviations from the mean}}{\text{number of figures in the list}}$$

For example, for 7, 3, 10, 4 the mean is 24/4 = 6 and hence the deviations from the mean are 1, 3, 4, 2. Therefore the mean deviation is:

$$\frac{1 + 3 + 4 + 2}{4} = \frac{10}{4} = 2 \frac{1}{2}.$$

Median   The median of a set of numbers is the value of the middle number, when they are arranged in ascending order. Examples:

For 1, 2, 4, 7, 9 the median is 4.
2, 5, 8, 3, 1, 7, 6 becomes 1, 2, 3, 5, 6, 7, 8 and the median is 5.

If there is no middle number, the average of the two middle numbers is taken. For example:

For 1, 5, 7, 8, 9, 10 the median is $\dfrac{7+8}{2} = 7\frac{1}{2}$.

The median of a frequency distribution is found from the **cumulative frequency** graph or **Ogive**.

Mediator   The mediator of a line is the line that bisects it at right angle.

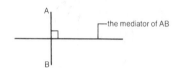

Member   A number, letter, symbol, etc., that belongs to a **set** is said to be a member of the set. For example:

   2 is a member of the set of even numbers
   a is a member of the set of vowels
but 4 is not a member of the set of odd numbers

The mathematical symbols used for 'is a member of' and 'is not a member of' are ϵ and $\notin$ respectively. For example:

   $2 \in \{\text{prime numbers}\}$
   $\text{s} \in \{\text{letters of alphabet}\}$
   $\text{o} \notin \{\text{vowels}\}$
   $x \notin \{\beta, \gamma, \pi, \rho\}$

Mensuration The measuring of geometrical quantities, for example, lengths, areas and volumes.

Meridian A meridian is a **great circle** drawn on the earth's surface which passes through the north and south poles. For example, all **longitude lines** may be referred to as meridians.

Metre A metre is a unit of length (just over thirty-nine inches). One metre = 100 centimetres (1 m = 100 cm).

Metric A system of units of measurement in which the fundamental units of length, mass, etc., are divided and compounded by factors of ten. For example, 1 kilogram = 1000 grams, 1 metre = 10 decimetres.

The prefixes kilo-, hecto-, deca-, deci-, centi-, milli-, denote multiples of 10^3, 10^2, 10, 1/10, 1/100, 1/1000.

Mile A unit of length. One mile = 5280 feet or 1760 yards. One mile is approximately 1.6 km.

Millimetre   A unit of length. There are ten millimetres in one centimetre. One millimetre equals 0.1 cm, 10 mm = 1 cm.

Million   One thousand thousands. One million equals 1 000 000.

Minimum   The graph shown is said to have a minimum point at the point P.

The **tangent** drawn to the curve at this point would have zero **gradient**. Just to the right of p, the gradient is positive and just to the left, it is negative.

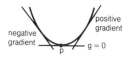

Minor 1) The shorter of the two axes of symmetry of an ellipse is called the minor axis. The other is called the **major** axis. For an ellipse with equation:

$$\frac{x^2}{a^2} + \frac{y^2}{b^2} = 1 \ (a > b)$$

the minor axis has length 2b.

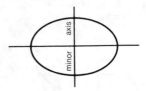

2) The shaded part of the diagram below, shows a minor **sector** of the circle. The **arc** ACB is called the minor arc of the circle.

Minute 1) A unit of time. There are sixty seconds in one minute, and sixty minutes in one hour.

2) A unit of angular measure. There are sixty minutes in one **degree** ($60' = 1°$).

Mixed number A mixed number is the sum of a whole number and a fraction: $2\frac{1}{2}$, $3\frac{2}{5}$, $-6\frac{7}{8}$ are all mixed numbers.

A mixed number can be changed into an **improper fraction** as follows:

$$2\tfrac{5}{7} = \frac{(2 \times 7) + 5}{7} = \frac{19}{7} \text{ or } 3\tfrac{9}{10} = \frac{(3 \times 10) + 9}{10} = \frac{39}{10}.$$

Mode The mode of a set of numbers is the number that occurs most often in the set. For example:

1, 2, 3, 3, 4, 6, 9 mode is 3
3, 4, 4, 4, 7, 7, 8 mode is 4
2, 2, 3, 5, 6, 9, 9, this set has two modes, 2 and 9

score on die	1	2	3	4	5	6
number of throws	5	6	2	4	3	5

mode is 2, since 2 was thrown most often (6 times)

In a **grouped** frequency distribution the group that contains most members is called the modal group or modal class. For example:

height of children (cm)	50–54	55–59	60–64	65–69
number of children	7	2	4	3

The modal group is 50–54 cms.

Modulo arithmetic In arithmetic modulo n, all numbers are represented by their remainder when divided by n. In modulo 8, $6 + 5 = 3$, $4 \times 3 = 4$, etc.

This type of arithmetic can be modelled by moving round a clockface.

$6 + 5 \equiv$ start at 6, then move on 5 places, to finish at 3.

In arithmetic modulo p, where p is prime, equations of the form $ax + b = c$ have unique solutions.

Modulus 1) The modulus of a **real number** x, denoted by $|x|$ is the positive value of x, regardless of its sign. For example: $|3.5| = 3.5$, $|-7| = 7$.

2) The modulus of a **complex number** is its distance from the origin when represented on the **Argand diagram**.

142

The modulus of the complex number $a + bi$, denoted by $|a + bi|$, is $\sqrt{a^2 + b^2}$.

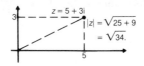

Multiple The number x is a multiple of the number y if y divides x exactly. For example, 5, 10, 15, 20, 25 . . . are all multiples of 5.

Multiplication One of the fundamental operations of arithmetic. Multiplication is associated with repeated addition. For example,
$5 \times 3 = 5 + 5 + 5 = 15$.

Analogous operations on abstract mathematical systems are sometimes known as multiplication. For example:

$$\text{matrix multiplication} \begin{pmatrix} 3 & 1 \\ 4 & 2 \end{pmatrix} \times \begin{pmatrix} 1 & 2 \\ 3 & 4 \end{pmatrix}$$

$$= \begin{pmatrix} 6 & 10 \\ 10 & 16 \end{pmatrix}$$

Natural The set of positive whole numbers (counting numbers) are called the natural numbers.

The set of natural numbers {1, 2, 3, 4, . . .} is often denoted by ℕ.

Natural **logarithms** are logarithms to the base of **e**.

Nautical mile A nautical mile (or sea mile) is a distance of 6080 feet, and is equivalent to one **minute** of arc measured along a **great circle** on the earth's surface. Hence 60 nm of great circle subtends an angle of 1° at the centre of the earth.

Negative A negative number is a number whose value is less than 0. For example, −2.5, −724, −0.176, −¹⁄₁₀ are all negative, 0, 3.5, 176, 0.001 are not negative.

Net A surface which can be folded into a solid.

Each of the shapes above is said to form a net for a **cube**, that is, if they were cut out, could be folded up to form a cube.

net of a
square-based
pyramid

net of a
tetrahedron

Newton A unit of force named after the famous English mathematician. A resultant force of 1 Newton acting on a body of mass 1 kg produces an **acceleration** of 1 metre per second per second.

Newton's method A step-by-step, **iterative** method of finding the roots of an equation of the form: $f(x) = 0$.

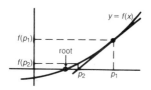

$y = f(x)$

$f(p_1)$

root

$f(p_2)$

p_2 p_1

If p_1 is an approximation to the root, then p_2, the **intercept** of the **tangent** to f at $x = p_1$ with the x-axis, is a better approximation.

Node A node is a point on a network to which one or more **arcs** lead.

nodes

Nonagon A nonagon is a nine-sided **polygon**. A regular nonagon has all its sides equal, and each of its interior angles measure 140°.

Normal The normal to a curve at any point is the line perpendicular to the **tangent** at that point.

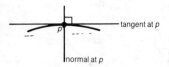

tangent at p

normal at p

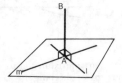

A line (or **vector**) is said to be normal to a plane if it is perpendicular to all lines belonging to the plane. l and m are lines belonging to the plane. AB is perpendicular to l and m and hence to the plane.

Notation   Symbolism used in mathematics. Symbols representing quantities, operations, relations, etc., together with the conventions regarding their use. For example:

The notation 5! means $5 \times 4 \times 3 \times 2 \times 1$.

The notation $\sum_{1}^{4} i$ means $1 + 2 + 3 + 4$.

Numeral   Numerals are symbols used to denote numbers. For example:

Arabic numerals 0, 1, 2, 3, 4, 5, 6, 7, 8, 9
Roman numerals I, V, X, L, C, D, M

Numerator   The top part of a fraction. For example, for $^{17}/_4$ the numerator is 17, for $^6/_{13}$ the numerator is 6.

Obtuse   An obtuse angle is greater than 90° but less than 180°.

this is an obtuse angle

147

Octagon An octagon is a **polygon** having eight sides.

A regular octagon has all its sides equal and all its interior angles are 135°.

Octahedron A **polyhedron** having eight faces.

A regular octahedron is made from eight equilateral triangles.

Odd An odd number cannot be divided exactly by two. For example, 7, 19, −107 are odd numbers.

If n is a whole number then $2n + 1$ is always odd.

Ogive An ogive is another name for a **cumulative frequency** curve.

Ogives are often shaped as in the diagram below.

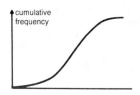

Operation An operation is a way of combining together members of a set.

Binary operations combine two members of a set to produce a third member as a result.

The four basic operations of arithmetic – addition, subtraction, multiplication and division – are examples of binary operations on the set of real numbers.

The study of abstract operations and their properties forms an important branch of modern algebra.

Order 1) The order of a **matrix** is the number of **rows** and **columns** of the matrix.

$A = \begin{pmatrix} 3 & 1 & 6 \\ 2 & 5 & 8 \end{pmatrix}$ Matrix A has order 2 × 3. This means it has two rows and three columns.

2) The order of a **node** in a network is the number of arcs that lead to that node.

node order 1
node order 3

3) A figure which maps to itself by reflections about *n* lines has line **symmetry** of order *n*.

A figure which maps to itself *n* ways by rotations about some centre 0 has rotational **symmetry** of order *n* about 0.

rotational symmetry of order 3

A square has line symmetry of order 4.

4) A **group** with *n* elements is said to be a group of order *n*.

Ordered pair A pair of numbers whose values and order are significant is termed an ordered pair. Used in particular to denote a pair of Cartesian coordinates specifying a point in a plane.

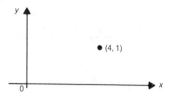

the ordered pairs (1, 4) and (4, 1)
specify different points

Ordered pairs such as (H, T), (H, H) etc., may also be used to specify the results of trials, for example, tossing two coins.

Ordinate The y-coordinate, or distance from the horizontal axis, of a point referred to a system of rectangular coordinates.

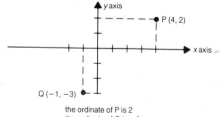

the ordinate of P is 2
the ordinate of Q is -3

Origin The origin is the point where the x and y axes cross. It has coordinates $(0, 0)$ and the coordinates of all other points are measured relative to the origin.

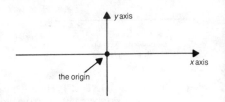

Orthocentre The orthocentre of a triangle is the point where the **altitudes** meet.

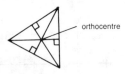

Parabola The **locus** of a point which moves so that it is equidistant from a fixed point (called the focus) and a fixed line (the directrix) is called a parabola.

152

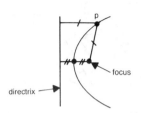

Any equation of the form $y = ax^2 + bx + c$ or $y^2 = 4ax$ will give a parabolic shape. The path of a **projectile** is approximately parabolic.

Parallel Lines in a plane that never meet no matter how far they are extended are called parallel. Parallel lines are always the same distance apart.

A pair of railway lines are parallel.

Parallelepiped A polyhedron whose faces are

all parallelograms is called a parallelepiped.

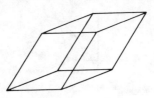

Parallelogram   A quadrilateral with its opposite sides equal (and parallel).

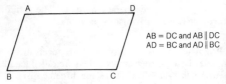

AB = DC and AB ∥ DC
AD = BC and AD ∥ BC

All parallelograms have the following properties:

1) diagonals bisect each other, and the whole shape
2) opposite angles are equal
3) rotational symmetry of order 2.

Parameter   In the equation $y = mx + c$, the values m and c are called parameters and specify the characteristics of the straight line represented by the equation. (*m* represents the **gradient**, and *c* the **intercept** on the *y*-axis.)

The parametric equation of the **parabola** $y^2 = 4x$ is $x = t^2$, $y = 2t$, each value of the parameter t gives a point on the curve.

Pascal   Pascal's triangle is a pattern of numbers:

$$
\begin{array}{ccccccccccc}
 & & & & & 1 & & & & & \\
 & & & & 1 & & 1 & & & & \\
 & & & 1 & & 2 & & 1 & & & \\
 & & 1 & & 3 & & 3 & & 1 & & \\
 & 1 & & 4 & & 6 & & 4 & & 1 & \\
1 & & 5 & & 10 & & 10 & & 5 & & 1
\end{array}
$$

Each number is the sum of the two numbers directly above it.

Penny   A unit of currency, one hundred pennies (pence) = one pound (100p = £1).

Pentagon   A five-sided **polygon**. A regular pentagon has all its sides equal and each of its interior angles is 108°.

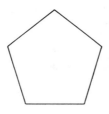

Percentage A method of relating a fraction of a given quantity to the whole in parts per hundred. For example:

30% of something is thirty parts per hundred.

Hence $30\% \equiv \dfrac{30}{100}$ (note % means 'per cent')

Percentage changes in a quantity are calculated as follows:

$$\text{If original value} = A$$
$$\text{new value} = B$$
$$\text{per cent change} = \frac{|A - B|}{A} \times 100$$

Percentile The **range** of a frequency distribution is divided into one hundred parts by percentiles. Fifteen per cent of the data lies below the fifteenth percentile, forty per cent lies below the fortieth percentile, etc.

The **median** value of a distribution occurs at the fiftieth percentile.

Perfect A perfect number is one that is equal to the sum of its **factors** (excluding the number itself). For example:

$$6 = 1 + 2 + 3 \qquad 28 = 1 + 2 + 4 + 7 + 14$$

The third perfect number is 496.

Perimeter   The perimeter of a figure is the distance measured round the boundary of the figure.

the perimeter
of this triangle
is 3 + 4 + 5 = 12 units

the peremeter of
this shape is 20 units
(2 + 2 + 4 + 2 + 4 + 4)

The perimeter of a circle is also known as the **circumference**.

Period The graph of $y = \sin x$ repeats itself every $360°$. This value 360 is said to be the period of the function.

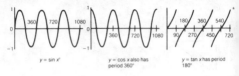

$y = \sin x°$

$y = \cos x$ also has period $360°$

$y = \tan x$ has period $180°$

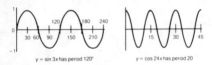

$y = \sin 3x$ has period $120°$

$y = \cos 24x$ has period 20

Permutation A permutation is an ordered arrangement of a set of numbers.

For example, all possible permutations of the numbers 1, 2, 3 are 1, 2, 3, 12, 13, 23, 21, 31, 32, 123, 132, 231, 213, 312 and 321.

There are $n!$ permutations of n numbers taken all at a time. For example:

1 number has 1 permutation
2 numbers have 2 permutations
3 numbers have 6 permutations
4 numbers have 24 permutations

There are $n!/(n - r)!$ permutations of n number taken r at a time.

Perpendicular   Two straight lines are said to be perpendicular if they meet at right angles.

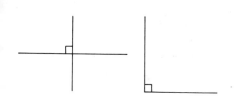

Two planes may also be perpendicular.

Pi   The symbol pi (π) is used to denote the ratio of the **circumference** of a circle to its **diameter**.

$$\pi = \frac{C}{D}$$

π is an **irrational** number: $\pi = 3.1415926535 \ldots$ we commonly use 3.14 or 22/7 as an (approximate) value for π.

π is used in calculating lengths, areas and volumes of circular figures and solids. For example:

Area of a circle is πR^2
Volume of a cylinder $\pi R^2 h$
Volume of a sphere $4/3 \pi R^3$

159

Pictogram A pictogram is a way of representing information (usually statistical) in the form of a picture.

pictogram showing
the vehicles passing
a checkpoint in one
minute

cars   🚗 🚗 🚗 🚗 🚗   (5)

buses   🚌 🚌   (2)

lorries   🚚 🚚 🚚   (3)

cycles   🚲 🚲   (2)

Pie chart A pie chart or circular diagram is another way of representing statistical information. The circle is divided into **sectors** whose areas represent the number in the set.

For example, for the information in the diagram above the angles would be:

Car $\dfrac{5}{12} \times 360° = 150°$

Bus $\left.\begin{matrix} \\ \\ \end{matrix}\right\}$ $\dfrac{2}{12} \times 360° = 60°$ each
Cycle

Lorry $\dfrac{3}{12} \times 360° = 90°$

Plan   The plan of a shape is the view of the shape when looking vertically downwards onto it.

The plan view of a cylinder is a circle.

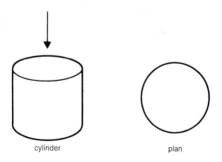

cylinder   plan

161

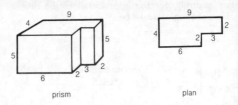

prism   plan

A map gives a plan view of a portion of country.

Plane A plane is a flat surface.

In **Cartesian coordinates** the equation of a plane is of the form $ax + by + cz = d$.

Point Points are often described by their **coordinates**. A point has position but no real size.

Polar coordinates Polar coordinates are a sys-

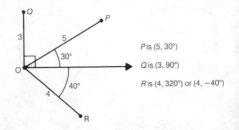

P is $(5, 30°)$

Q is $(3, 90°)$

R is $(4, 320°)$ or $(4, -40°)$

tem of specifying the position of points by their distance from a fixed point, and angle measured from a fixed line.

Polar coordinates are of the form (r, θ) as opposed to (x, y) of **Cartesian coordinates**.

Polygon   A polygon is a plane shape having many (three or more) sides.

Name of polygon	Number of sides	Sum of interior angles
Triangle	3	180°
Quadrilateral	4	360°
Pentagon	5	540°
Hexagon	6	720°
Heptagon	7	900°
Octagon	8	1080°
Decagon	10	1440°
Duodecagon	12	1800°
Icosogon	20	3240°

Regular polygons have all their sides and all their angles equal.

In general a polygon having *n* sides will have *n* interior angles that sum to $(n - 2) \times 180°$.

Polyhedron A solid shape having many faces, each of which is a **polygon**.

There are only five regular polyhedra: **tetrahedron**, **cube**, **octahedron**, **dodecahedron**, **icosahedron**.

Polynomial An algebraic expression containing only one **variable** (usually *x*). For example:

$3x^2 + 5x + 1$ is a quadratic polynomial
$y^3 + 7y + 2$ is a cubic polynomial

In the polynomial $5x^4 + 3x^2 + 7x - 6$, the number 5 (which multiplies x^4) is called the **coefficient** of x^4.

Population A statistical term referring to the set of items for which a certain characteristic is being measured. Populations may be finite or infinite.

Positive A positive number is one whose value is greater than 0. For example:

735, 3¼, 0.1 are all positive

but

0, −2, −3.5 are not positive

Pound 1) A unit of currency. £1 equals 100 pence.

2) A unit of weight. One pound equals 16 ounces (1 lb = 16 oz), 14 pounds equals 1 stone.

Power For example 2 to the power 4 (written 2^4) = $2 \times 2 \times 2 \times 2$ = 16, 3 to the power 2 (written 3^2) = 3×3 = 9.

Any number to the power 2 is said to be 'squared' any number to the power 3 is said to be 'cubed'.

Prime A prime number is one that has exactly two **factors**, 1 and the number itself. There are an infinite number of prime numbers. For example, 2, 3, 5, 7, 11, 13, 17, 19, etc.

Note: 2 is the only even prime number.

Prism A prism is a polyhedron having the same **cross section** throughout its length.

a cylinder is a prism whose cross section is a circle

a triangular prism

The volume of a prism is found by multiplying the cross sectional area by the length of the solid.

Probability Probability is a measure of how likely an event is. Its value lies between 0 and 1. For example:

Probability of the sun rising in the east = 1.
Probability that Sunday is the day after Monday = 0.
Probability of throwing a head with a coin = 0.5.

In situations where several equally likely outcomes are possible, the probability of a particular event can be measured by:

$$\frac{\text{number of events favourable to the outcome}}{\text{total number of possible events}}$$

Product The product of two or more numbers is the result of multiplying them together. For example:

Product of 2, 3, 4 is $2 \times 3 \times 4 = 24$.
Product of 6 and 11 is $6 \times 11 = 66$.

The product of the matrices

$$\begin{pmatrix} 2 & 3 \\ 4 & 1 \end{pmatrix} \text{ and } \begin{pmatrix} 3 & 5 \\ 1 & 2 \end{pmatrix} \text{ is } \begin{pmatrix} 9 & 16 \\ 13 & 22 \end{pmatrix}.$$

Progression A **sequence** of numbers each of which is related to the previous one by a simple connection. For example:

2, 5, 8, 11, is an **arithmetic** progression
2, 6, 18, 54 is a **geometric** progression

Projectile A particle thrown into the air is known as a projectile. The path of a projectile is roughly parabolic.

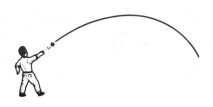

167

Projection If a light was shone vertically downwards onto the plane PQRS, the line AB would cast its shadow along the line CD. CD is said to be the projection of AB onto the plane.

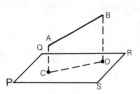

Proof The logical argument used to establish the truth of a statement.

Proper A proper fraction is one whose **numerator** is less than its **denominator**. For example, ³⁄ is a proper fraction but ⁷⁄3 is not.

A proper **subset** A of a set B is a subset which does not contain all the elements of B. That is A ∩ B = A but A ∩ B ≠ B. For example:

{1, 2, 3} is a proper subset of {1, 2, 3, 4, 5, 6} but {1, 2, 3, 4, 5, 6} is not a proper subset of {1, 2, 3, 4, 5, 6}.

Proportion Two sets of numbers are said to be in proportion when the **ratio** between corresponding members of the two sets is constant. For example:

{1, 2, 5, 8, 10} and {2, 4, 10, 16, 20} are in proportion, since $1:2 = 2:4 = 5:10 = 8:16 = 10:20$. (That is, each member of set two is twice the corresponding member of set one.)

Two sets are inversely proportional when the members of one set are proportional to the reciprocals of the members of the other set. For example:

{1, 2, 3, 4, 6} is inversely proportional to {12, 6, 4, 3, 2}

In this case the product of the corresponding elements is constant, i.e.,
$$1 \times 12 = 2 \times 6 = 3 \times 4 = 4 \times 3 = 6 \times 2.$$

Protractor A device used for measuring angles.

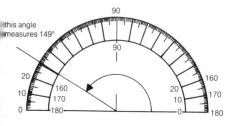

Pyramid A pyramid is a **polyhedron** with a **polygon** for its base the other faces being triangles with a common **vertex**.

square-based pyramid

triangular pyramid

hexagonal pyramid

The volume of a pyramid is given by:

V = ⅓ (area of base) × (perpendicular height)

V = ⅓ AH

A **cone** is similar to a pyramid, the formula for its volume resembles that for a pyramid:

$$V = \frac{1}{3} \pi r^2 h.$$

Pythagoras Pythagoras' Theorem states that in any right-angled triangle, the area of the square on the **hypotenuse** is equal to the sum of the areas of the squares on the other sides.

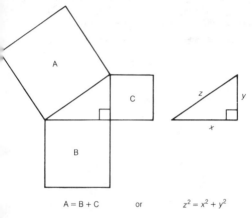

$$A = B + C \qquad \text{or} \qquad z^2 = x^2 + y^2$$

For example, the following triples of numbers all correspond to the lengths of sides in right-angled triangles:

3, 4, 5 $(3^2 + 4^2 = 9 + 16 = 25 = 5^2)$
6, 8, 10
5, 12, 13
7, 24, 25

8, 15, 17
9, 20, 21
20, 21, 29, etc

Pythagoras' Theorem states a fundamental property of space, and is frequently used to find unknown lengths in geometrical configurations.

Quadrant 1) A quarter of a circle is called a quadrant.

2) One of the four parts that the plane is divided into by the x and y axes:

	y axis	
second quadrant		first quadrant
		—x axis
third quadrant		fourth quadrant

Quadratic A quadratic **polynomial** is one of the second **degree**. For example, $3x^2 + 5x - 11$.

An equation that can be written in the form $ax^2 + bx + c = 0$ with $a \neq 0$ is called a quadratic equation.

Such equations may have two distinct, two coincident, or 0 real solutions depending upon the value of the **discriminant** of the equation.

The graph of quadratic function $y = ax^2 + bx + c$ is **parabola**-shaped.

The solutions of a general quadratic equation $ax^2 + bx + c = 0$ can be found by using the formula:

$$x = \frac{-b \pm \sqrt{b^2 - 4ac}}{2a}$$

Quadrilateral A quadrilateral is a polygon having four sides. **Square, rectangle, rhombus, parallelogram, kite, trapezium** are all special kinds of quadrilaterals.

examples of
quadrilaterals

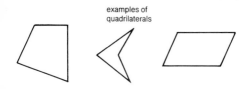

Quarter The name given to one fourth part of a shape or quantity.

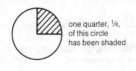

one quarter, ¼, of this circle has been shaded

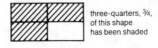

three-quarters, ¾, of this shape has been shaded

One quarter of 12 is 3, since $12 \div 4 = 3$ or $\frac{1}{4} \times 12 = 3$.

Quartic A quartic polynomial is a **polynomial** containing powers of x up to x^4. For example, $x^4 + 2x^3 - 7x + 1$.

$x^4 + 7x^3 - 2x + 3 = 0$ is a quartic equation.

Quartile A statistical measure used in connection with **cumulative frequency**. The first or lower quartile of a set of data is the value below which one quarter of the data lies. The second quartile (more commonly known as the **median**) is the value below which one half of the data lies, and the third or upper quartile is the value below which three-quarters of the data lies. They correspond to the twenty-fifth, fiftieth and seventy-fifth **percentiles**.

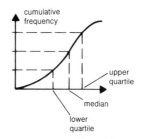

In a list of n numbers the lower quartile is the ¼ $(n + 1)$th, the median is the ½$(n + 1)$th, and the upper quartile is the ¾ $(n + 1)$th. For example:

1, 2, 3, 5, 6, 6, 7, 7, 9, 9, 10

lower quartile   median   upper quartile

Lower quartile = ¼ (11 + 1) = third value
Median = ½ (11 + 1) = sixth value
Upper quartile = ¾ (11 + 1) = ninth value

Quintic A quintic **polynomial** is a polynomial containing powers of x up to x^5. For example, $x^5 + 3x^4 - 2x^2 + 7$.

$x^5 + 3x^4 + 2x^3 + 6 = 0$ is a quintic equation.

Quotient The quotient is the result obtained when a division is performed. Examples:

$$\begin{array}{r} 29 \text{ (r16)} \\ 21\overline{)625} \end{array}$$ The quotient is 29 and **remainder** 16.

$$\begin{array}{r} x + 4 \\ x + 1\overline{\smash{\big)}\ x^2 + 5x + 4} \\ \underline{x^2 + x} \\ 4x + 4 \\ \underline{4x + 4} \\ 0 \end{array}$$ The quotient is $x + 4$ with no remainder.

Radian The angle **subtended** at the centre of a circle by a minor **arc** of length equal to the radius of the circle is called a radian.

One Radian $\simeq 57°$ and in a full circle there are 2π or 6.28 (approx) radians. One degree equals $\dfrac{\pi}{180}$ radians.

Radius The radius of a circle is the distance from the centre of the circle to any point on the **circumference**.

Random A **sample** taken from a population is said to be random if every member of the population has an equal chance of being chosen.

A random number is a number chosen from the set of values 0, 1, 2, 3, 4 . . . 9, where each value has the same chance of being chosen.

Range 1) A statistical term referring to the difference between the smallest and largest members of a set of numbers. For example, for 2, 3, 4, 7, 9, 10, 12, 15, the range is $15 - 2 = 13$.

2) The range of a **function** is the set of **images** of the members of the **domain** of the function. For example:
for the function $x \rightarrow 3x + 1$ with domain {1, 2, 3, 4}, the range is {4, 7, 10, 13}.

Rank When a set of numbers or quantities are arranged in ascending order, the position of any element in the list is called its rank. For example:

For the numbers	4	8	5	6	2	6	0	12
Rank order is	0	2	4	5	6	6	8	12
	↓	↓	↓	↓	↓	↓	↓	↓
Rank	1	2	3	4	5½	5½	7	8

Ratio A method of comparing two quantities measured in similar units by considering the **quotient** of the two quantities. For example:

The ratio of £2 to 50p is written: $200:50=4:1$.

Note: ratios are simplified similarly to fractions.

Rational A rational number is any number that can be expressed in the form of a fraction, i.e., in the form a/b where a and b are whole numbers, and $b \neq 0$. Examples:

$$2½ = \frac{5}{2} \quad 1.3 = \frac{13}{10} \quad 7 = \frac{7}{1} \quad \sqrt{0.81} = 0.9 = \frac{9}{10}$$

The symbol for the set of all rational numbers is Q. Rational numbers can always be written as **terminating** or **recurring** decimals.

Numbers that cannot be written as a fraction are called **irrational** numbers.

Rationalize To rationalize an algebraic expression is to remove operations of the form $\sqrt{}$, $\sqrt[3]{}$, etc. without changing the value of the expression. Examples:

$\sqrt{x-2} = x$ rationalizes into $x - 2 = x^2$ or $x^2 - x + 2 = 0$

$\sqrt[3]{x^2 + 1} = 2$ rationalizes into $x^2 + 1 = 8$ or $x^2 - 7 = 0$

In the example below the denominator of the fraction has been rationalized:

$$\frac{1}{\sqrt{x} + y} = \frac{\sqrt{x} - y}{x - y^2}$$

To obtain this result the numerator and denominator of the fraction are both multiplied by $\sqrt{x} - y$.

Ray A ray is a straight line extending from a point, called the origin of the ray. Another name for a ray is a half-line.

The diagram above shows three rays all with 0 as common origin.

Real numbers The set of real numbers is the **union** of the sets of **rational** and **irrational** numbers.

To each point on the continuous real number line there corresponds a real number.

$$-3 \quad -2 \quad -1 \quad 0 \quad 1 \quad 2 \quad 3 \quad 4$$

$$-1.5 \quad 0.\dot{3} \quad \sqrt{2} \quad e \quad \pi$$

$-1.5, 0, 0.\dot{3}, \sqrt{2}, e, \pi$ are all real numbers

Reciprocal The reciprocal of a non zero number x is the number $1/x$. For example, the reciprocals of 5, 3/4 and 0.4 are 1/5, 1/3/4 = 4/3, 1/0.4, respectively.

The product of any number and its reciprocal is equal to 1.

Rectangle A rectangle is a **quadrilateral**, all of whose angles are equal to 90°. Opposite pairs of sides are of equal length. A rectangle has two lines of **symmetry** (shown in the diagram below) and rotational symmetry of order 2. The **diagonals** of a rectangle bisect each other and the figure.

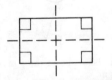

Rectangle number A rectangle number is one having more than two **factors**. These numbers can be represented by rectangular configurations of dots. Examples:

$$12 = 2 \times 6 \qquad \text{or} \quad 3 \times 4$$

. . . .
. . . .
                     . . . .

$$20 = 2 \times 10 \qquad\qquad \text{or} \quad 4 \times 5$$

. . . .
. . . .
                             . . . . .
                             . . . . .

The set of rectangle numbers is {4, 6, 8, 9, 10, 12, 14, 15, 16 . . .}. The set of **prime numbers** is {2, 3, 5, 7, 11 . . .}. Notice that the **union** of these two disjoint sets gives the set of all **natural numbers** (apart from 1 which is unique in having just one factor).

Rectangular 1) A rectangular **prism** is a prism whose cross section is a rectangle (same as a cuboid).

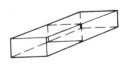

2) The x and y axes are said to be rectangular axes, being perpendicular to each other.

3) A rectangular **hyperbola** is one whose **asymptotes** are the x and y axes. The equation of a rectangular hyperbola is $xy = c^2$.

Rectilinear Rectilinear motion is motion along a straight line.

Recurring A recurring decimal contains an infinitely repeating block of decimal digits. Examples:

$\frac{1}{6} = 0.16666\ldots$ written as $0.1\dot{6}$
$\frac{2}{11} = 0.181818\ldots$ written as $0.\dot{1}\dot{8}$

All recurring decimals represent **rational numbers**. Fractions with denominators containing prime **factors** other than 2 or 5 will recur if written in decimal form.

Re-entrant A **polygon** is said to be re-entrant if one or more of its angles is greater than 180° (i.e., **reflex**).

this angle is greater than 180°

Non–re-entrant polygons are called **convex**.

Reflection A reflection is a geometrical transformation of the plane in which points are mapped to **images** by folding along a mirror line.

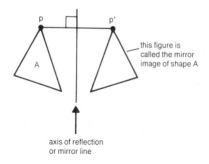

p

p′

A

this figure is called the mirror image of shape A

axis of reflection or mirror line

The object and image are **congruent**. A line joining any point P to its image P′ will always be perpendicular to the mirror line and the distances Px and P′x will be equal (see diagram above).

A **matrix** can be used to represent some reflections. Some of the simpler ones are given below:

Reflection in:

$$x \text{ axis } (y = 0) \quad \begin{pmatrix} 1 & 0 \\ 0 & -1 \end{pmatrix}$$

$$y \text{ axis } (x = 0) \quad \begin{pmatrix} -1 & 0 \\ 0 & 1 \end{pmatrix}$$

$$\text{the line } y = x \quad \begin{pmatrix} 0 & 1 \\ 1 & 0 \end{pmatrix}$$

$$\text{the line } y = -x \quad \begin{pmatrix} 0 & -1 \\ -1 & 0 \end{pmatrix}$$

Note: points on the mirror line do not change their position after the reflection.

Reflex A reflex angle is one that is greater than 180° but less than 360°.

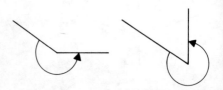

184

Reflexive A **relation** defined on a set is said to be reflexive if every member of the set is related to itself.

Region A network divides the plane into regions. Regions are spaces enclosed by the **arcs** of the network.

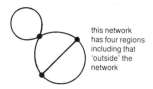

this network has four regions including that 'outside' the network

The graph of a **linear** relation, for example, $x + y = 5$, divides the plane into two regions.

Each region can be specified by an **ordering**. In this case $x + y < 5$ and $x + y > 5$.

Regular A regular **polygon** is one having all its sides equal, and all its interior angles equal.

A regular triangle is called an **equilateral** triangle. A regular quadrilateral is a **square**.

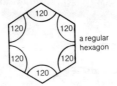

a regular hexagon

A regular **polyhedron** is one in which all the faces are the same regular polygon.

a regular **tetrahedron** is made up of four equilateral triangles

Relation A relation is a connection between the members of a set, or between the members of two sets one to another. For example, 'is older than' on a set of people; 'is parallel to' on a set of lines.

Relations between two sets of numbers are of particular importance. For example:

$x \to x^2 + 1$ between the set of all real numbers and $[1, \infty[$.

Remainder When 14 is divided by 4, the **quotient** is 3 (because $4 \times 3 = 12$) and there is a remainder of 2.

In any division process where the **dividend** is not a **multiple** of the **divisor**, there will be a remainder involved. For example:

$362 \div 25 = 14$, remainder 12.

Note that $362 = (14 \times 25) + 12$, i.e., dividend = (quotient × divisor) + remainder.

The remainder theorem for **polynomials** states that when a polynomial P(x) is divided by $(x - a)$ the remainder is P(a). For example:

when P(x) = $x^2 + 7x + 2$ is divided by $(x + 3)$ the remainder is P(-3) = $(-3)^2 - (7 \times 3) + 2 = -10$.

Notice that if $(x - a)$ is a **factor** of P(x), then P(a) = 0. For example:

$(x - 2)$ is a factor of P(x) = $x^2 + 4x - 12$ and P(2) = 0.

Repeated root If a root of an equation occurs more than once it is said to be a repeated or multiple root of the equation. For example, the expression:

$x^3 - 7x^2 + 16x - 12 = (x - 2)(x - 2)(x - 3)$.

Hence the equation $x^3 - 7x^2 + 16x - 12 = 0$ has three roots 2, 2 and 3. 2 is a repeated root.

Residue When the numbers 8, 23, 38 etc are divided by 5, they leave a **remainder** of 3, similarly 2, 17, 42, etc. leave a remainder of 2. We say that 8, 23, 38 all belong to the residue class $\bar{3}$ (**modulo** 5) and 2, 17, 42 belong to the residue class $\bar{2}$ (modulo 5). For modulo 5 there are five residue classes, denoted by $\bar{0}$, $\bar{1}$, $\bar{2}$, $\bar{3}$, $\bar{4}$ according to whether the remainder is 0, 1, 2, 3, 4.

Resolve To resolve a **vector** in any particular direction is to find the **component** of the vector in the given direction.

The resolved part of the vector **d** in the x direction is vector **OB** and in the y direction is **OA**

$$|OB| = |d| \times \cos 40°$$
$$|OA| = |d| \times \sin 40°$$

The resolved part of the vector **r** in the direction **AB** is $|\mathbf{r}| \times \cos \theta$ where θ is the angle between the directions of **r** and **AB**.

Resultant   The resultant of two or more **vectors**, is the single vector that would have the same effect as all the others combined.

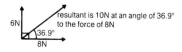

6N

resultant is 10N at an angle of 36.9° to the force of 8N

36.9°

8N

Revolve   To **rotate** about an axis or point.

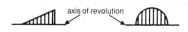

axis of revolution

If the shapes are rotated 360° about the axes marked, solids of revolution will be formed. The triangle will produce a cone and the semi-circle will produce a sphere.

Rhombus A rhombus is a **parallelogram** with all of its sides equal. A rhombus has two lines of **symmetry** and rotational symmetry of **order** 2. The diagonals of a rhombus bisect each other at right angles, and bisect the figure.

Right angle An angle equal to 90°, or a quarter of a full turn.

common system of marking a right angle

Rolle's theorem If a **continuous** curve crosses the x-axis at two points, A and B, and it is possible to draw a **tangent** to the curve at all points between A and B, then there will be at least one point on the

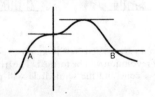

190

curve where a tangent can be drawn parallel to the x-axis.

Roman numerals A system of writing the **natural numbers** used by the Romans:

$$I = 1 \quad V = 5 \quad X = 10 \quad L = 50$$
$$C = 100 \quad D = 500 \quad M = 1000$$

All other numbers are made using combinations of the letters above:

4 = IV	(1 before 5)
6 = VI	(1 after 5)
27 = XXVII	(7 after 20)
90 = XC	(10 before 100)
1982 = MCMLXXXII	

Root A root of an equation is a value which when substituted into the equation in place of the unknown, will make the equation true. For example:

the roots of $x^2 - 7x + 12 = 0$ are 3 and 4
since $3^2 - 7 \times 3 + 12 = 0$
and $4^2 - 7 \times 4 + 12 = 0$

2 is not a root because $2^2 - 7 \times 2 + 12 \neq 0$.

A number is said to be a root of another if by taking **powers** of the number it is possible to make the other. Examples:

5 is the square root of 25 since 5×5 or $5^2 = 25$
2 is the cube root of 8 since $2 \times 2 \times 2$ or $2^3 = 8$
3 is the fifth root of 243 since $3^5 = 243$

Rotation A rotation is a geometrical transformation where every point turns through the same angle about a single point called the centre of rotation. The centre of rotation is the only point which does not change its position after the rotation.

the rectangle has rotated 90° about the point 0

the triangle has rotated 90° about the point 0

To distinguish between **clockwise** and anticlockwise rotations, + and − signs are used, − for clockwise and + for anticlockwise. For example, +60° means a rotation of 60° anticlockwise, −90° means a rotation of 90° clockwise.

Note that +90° = −270° and −100° = +260°, etc.

Some rotations can be expressed in **matrix form**. For example:

$$+90° \text{ about } (0, 0) = \begin{pmatrix} 0 & -1 \\ 1 & 0 \end{pmatrix}$$

$$-90° \text{ about } (0, 0) = \begin{pmatrix} 0 & 1 \\ -1 & 0 \end{pmatrix}$$

Rounding off Rounding off is a way of rewriting a number with fewer non-zero digits. For example:

Round off £16.25 to the nearest £1 – £16
Round off 275 m to the nearest 100 m – 300 m
Round off 25471 to the nearest 1000 – 25000

When the first digit to be ignored is a 0, 1, 2, 3, 4, we 'round down'; when it is a 5, 6, 7, 8 or 9 we 'round up'. For example:

273$\underline{8}$1	to the nearest 100 is 27400	(round up)
42.$\underline{7}$3	to the nearest 1 is 43	(round up)
6.1$\underline{2}$38	to the nearest $\frac{1}{10}$ is 6.1	(round down)
86$\underline{5}$	to the nearest 10 is 870	(round up)

(The first digit being ignored is underlined.)

Row A list of numbers or letters written horizontally is known as a row. For example:

1 2 3 4 5 6
(2, 5) **coordinates** are written as a row

$\begin{pmatrix} 1 & 2 & 5 \\ 3 & 5 & 6 \end{pmatrix}$ this **matrix** has two rows

$\begin{pmatrix} 1 & 7 & 6 & 4 & 2 \end{pmatrix}$ a matrix containing only one row is called a row matrix

Ruler A straight edge marked in linear units and used for measuring distances.

Sample A finite proportion of a **population** (which may be infinite). For example:

a sample from the set of **integers** could be 1, 3, 5, 7, 9, 11 or 1, 4, 9, 16, 25 or −1, −2, −3, −8, etc.

a sample from the letters in the alphabet could be {vowels} or {consonants}.

Satisfy 1) To fulfil the conditions of. For Example:

120° satisfies the requirements of being an **obtuse** angle.

2) A value or set of values is said to satisfy an equation if, when they are **substituted** into the equation, they make the equation balance. For example:

$x = 3$ satisfies $x^2 + 5 = 14$ since $3^2 + 5 = 14$ is true.

$x = 2$, $y = 3$ satisfy the equations $x + y = 5$ and $2x - y = 1$ since $2 + 3 = 5$ and $2 \times 2 - 3 = 1$ are both true.

Scalar A scalar is a quantity that has size only but not direction (as opposed to a **vector** which has both size and direction). For example, time, distance, mass, volume are all scalars.

Scale 1) A series of marks on a ruler, measuring cylinder, etc., used as an aid in measuring distances, volumes etc.

2) A scale drawing of an object is produced with all measurements in the same **ratio** with the corresponding measurements of the original.

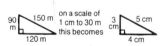

Scale factor When a shape is **enlarged**, the scale factor of the enlargement is defined as:

$$\frac{\text{distance between any two points on the image}}{\text{the corresponding distance on the object}}$$

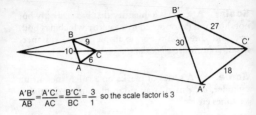

$$\frac{A'B'}{AB} = \frac{A'C'}{AC} = \frac{B'C'}{BC} = \frac{3}{1} \text{ so the scale factor is 3}$$

If the areas of the shapes were compared, the enlarged shape would have an area nine times larger than the original shape. We say that the area scale factor is nine.

If a solid shape is enlarged, for example, a cuboid 2 by 3 by 4 becomes a cuboid 4 by 6 by 8, we notice that all linear measurements are doubled, all areas are multiplied by four and the volume is eight times larger. We say:

linear scale factor $= 2$
area scale factor $= 4$ (note $4 = 2^2$)
volume scale factor $= 8$ (note $8 = 2^3$)

Scalene A scalene triangle is one where no two sides or angles are equal.

a scalene triangle may be right-angled

3 cm, 4 cm, 5 cm

Scatter diagram A diagram showing the **frequencies** with which joint values of variables occur.

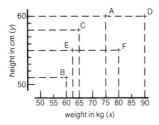

	weight (kg)	height (cm)
A	75	60
B	60	51
C	65	58
D	90	60
E	62	55
F	80	55

Secant A trigonometric function. In a right-angled triangle the secant of an angle is the ratio:

$$\frac{\text{hypotenuse}}{\text{adjacent side}}$$

in the triangle ABC

$$\sec \alpha = \frac{AC}{AB} = \frac{10}{8} = 1.25$$

The value of α can be determined by examining a table of values for the secant function $\alpha = 36.9°$.

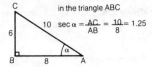

in the triangle DEF

$$\sec 40° = \frac{EF}{3.5}$$

From the table of values for the secant function $\sec 40° = 1.31$. Hence:

$$1.31 = \frac{EF}{3.5}$$

and EF = 1.31 × 3.5 = 4.59.

Segment A segment is part of a circle cut off by

198

chord. When the chord goes through the centre of the circle the two segments are called semi-circles.

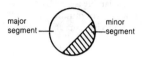

Self-inverse A member of a set is said to be self-inverse if it combines with itself to give the **identity** element of the set under a given operation. For example:

the **matrix** $\begin{pmatrix} -1 & 0 \\ 0 & -1 \end{pmatrix}$ is self inverse under matrix multiplication since

$$\begin{pmatrix} -1 & 0 \\ 0 & -1 \end{pmatrix} \times \begin{pmatrix} -1 & 0 \\ 0 & -1 \end{pmatrix} = \begin{pmatrix} 1 & 0 \\ 0 & 1 \end{pmatrix}.$$

For example, the element 2 is self inverse under addition **modulo** 4 since $2 + 2 = 0$.

For example:

*	a b c d
a	a b c d
b	b c d a
c	c d a b
d	d a b c

For this operation *, a is the identity element so c is self inverse since $c*c = a$.

Second 1) A unit of time. There are sixty seconds in one minute.

2) A unit of circular measure. Each 1° of angle can be subdivided into sixty minutes (1° = 60′) and each minute can be subdivided into 60 seconds (1′ = 60″).

Section A cross section of a solid shape is a view of the shape taken perpendicular to its axis of symmetry.

the cross section of a cone is a circle

the cross section of this prism is an equilateral triangle

Sector A sector is part of a circle enclosed between two *radii*.

minor sector

major sector

The area of a sector of $\theta° = \theta/360 \times \pi R^2$.

The arc length of a sector of $\theta°$ is $\theta/360 \times 2\pi R$. Where R is the radius of the circle.

Semi 1) A semi-circle is half a circle and is formed when a **chord** passes through the centre of a circle.

Area = $\frac{1}{2}\pi R^2$
circumference = πR

2) The semi-**interquartile range** of a set of numbers is defined as: $\frac{1}{2}$ (upper quartile − lower quartile). For example, 1, 3, 7, 8, 9, 10, 12, 13, 17, 20, 27.

Upper quartile = 17 $\Big\}$ semi-interquartile range
Lower quartile = 7 $\Big\}$ = $\frac{1}{2}$ (17 − 7) = 5.

Sense Sense is a word used in connection with transformations. If the shape transformed is 'flipped over' by the transformation we say a change of sense has occurred. Rotations, translations, enlargements and shears do not change the sense of a shape, but reflections always do.

Sequence A set of quantities, often numbers, that are ordered a_1, a_2, a_3, . . . so that each member of the sequence corresponds to a particular **natural number**. Sequences are sometimes defined by a formula, for example, $a_n = 2n^2 - 1$ yields the sequence 1, 7, 17, . . . Sequences can also be defined inductively by relating each term to its predecessors, for example, $a_n = 3a_{n-1} + 1$ and $a_1 = 1$ yields the sequence 1, 4, 13, 40. . .

Sequences can have a finite or infinite number of terms.

Series The sum of a finite or infinite number of terms. $a_1 + a_2 + a_3 + \ldots$

The study of series and associated properties relating to sums, **limits** and **convergence** is an important branch of pure mathematics.

The sum to n terms of the series of terms in an **arithmetic progression** $a + (a + d) + (a + 2d) + (a + 3d)$ . . . is:

$$\frac{n}{2}(2a + [n - 1]d).$$

The sum to n terms of the series of terms in a **geometric progression** $a + ar + ar^2 + ar^3$ . . . is:

$$\frac{a(1 - r^n)}{1 - r}.$$

Set   A collection of distinct objects or things.

A set comprises members or elements, of which there may be a finite or infinite number.

The symbol { } means 'the set of', and the symbol ϵ means 'is a member of'.

Sets may be defined by listing or describing elements, and are often denoted by a capital letter. For example:

E = {even numbers} and 4ϵE
V = {a, e, i, o, u} and iϵV

Sexagesimal   A system related to a base of sixty. For example, the common system of degree measure for angles is a sexagesimal system.

360 degrees = 1 revolution
60 minutes = 1 degree
60 seconds = 1 minute

Shear   A transformation in which a line (in 2-D), or a plane (in 3-D) remains fixed whilst all other points move, parallel to the fixed line or plane, a distance proportional to their distance from the line or plane.

The area of a closed figure in a plane remains **invariant** under a shear. Likewise the volume of a solid remains invariant under a shear.

Shears with invariant lines or planes through the origin of coordinates can be represented by **matrices**. For example:

$$\begin{pmatrix} 1 & 0 \\ 3 & 1 \end{pmatrix}$$

represents a shear in the x–y plane with the y axis being held fixed.

Various geometrical and spacial properties of figures can be explored through the use of the shearing transformation.

The area of a parallelogram can be related to the area of a rectangle by a shear.

Side A side of a polygon is one of the line segments forming the boundary of the figure. For example, a quadrilateral has four sides.

Sieve (of Eratosthenes) An ancient process for finding **prime numbers** by striking out from the list of natural numbers firstly the multiples of 2 (4, 6, 8 . . .), then 3, then 5, etc., thus leaving the prime numbers.

Sigma A letter of the Greek alphabet. Upper case Σ and lower case σ.

The former of these two symbols is often used to signify a summation. For example:

$$\sum_{1}^{5} a_i = a_1 + a_2 + a_3 + a_4 + a_5$$

$$\sum_{1}^{4} i^2 = 1^2 + 2^2 + 3^2 + 4^2 = 30$$

The latter lower case symbol is often used to signify the **standard deviation** of a distribution in statistics.

Sign Notation used to signify whether a quantity is positive or negative. Positive sign $+$, negative sign $-$.

Also used in relation to the symbols representing various mathematical operations and relations. For example, multiplication sign $\times$, equals sign $=$, square root sign $\sqrt{\ }$.

Significant In statistics a difference or deviation between observations and a proposed theory are called significant if they are unlikely to be the result of chance fluctuations.

The significance of a particular digit in a number is concerned with its relative size and importance in the number, and is dependent on the place value of the digit.

For example, in the number 39.08, 3 is the most significant digit, whilst 8 is the least significant.

Measurements are often **rounded off** to prescribed numbers of significant figures. For example, 27.058 can be rounded off thus:

30 to 1 significant figure
27 to 2 significant figures
27.1 to 3 significant figures, etc.

Similar Objects, plane or solid figures, are similar if they have the same shape.

When a shape is subject to an **enlargement**, the shape and its image are similar.

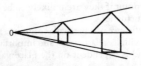

The corresponding angles contained in similar figures are equal, and the **ratio** of the lengths of corresponding sides of similar figures is constant.

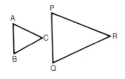

∠ BAC = ∠ QPR, ∠ ACB = ∠ PRQ, ∠ ABC = ∠ PQR
AB : PQ = BC : QR = CA : RP

Simple closed curve A curve that can be drawn with a single sweep of a pencil, with the end of the curve joined to the beginning, and such that the curve does not cross itself at any point.

Simple Harmonic Motion (S.H.M.) The motion of a body that oscillates about a fixed point, its **acceleration** being directed towards the fixed point, and proportional to its distance from the point.

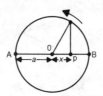

Motions of this type can be modelled by the projection onto a diameter of a particle moving with constant angular velocity ω round a circle.

The time for a whole oscillation is called the **period** of the motion. The extreme distance from the centre is called the **amplitude**.

Acceleration $= -\omega^2 x$
Distance from centre $x = a \cos(\omega t + \propto)$
Period $= 2\pi/\omega$

Simplify To contract an expression by the use of algebraic or arithmetical techniques. For example, $6/9 = 2/3$ or $3x + 2x + 4x = 9x$.

Simpson's Rule A method for finding the approximate values of definite **integrals**.

The method derives from dividing the interval over which the integral is defined into strips of equal width bounded above by **quadratic** curves, which are taken to approximate closely to the function being considered.

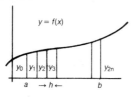

The value of $\displaystyle\int_a^b f(x)\,dx$

given by Simpson's Rule with $2n$ strips of width h is:

$$\frac{1}{3}h[(y_0 + y_{2n}) + 4(y_1 + y_3 + \ldots) + 2(y_2 + y_4 + \ldots)]$$

For example, using the rule with four strips of unit width:

$$\int_1^5 x^3 dx \simeq \frac{1}{3} \times 1 \times [(1 + 125) + 4(8 + 64) + 2(27)]$$

$$\simeq \frac{468}{3}$$

$$\simeq 156$$

Simultaneous equations A system of several equations with several unknowns (often linear). Examples:

$$3x + y = 17 \quad \text{or} \quad x^2y + y^2 = 26$$
$$2x - 3y = 12 \quad\quad x + y = 6$$

or
$$3p + 2q + 5r = 6$$
$$2p + 3q + 4r = 7$$
$$4p + 3q - 2r = 8$$

Linear simultaneous equations may be solved in a variety of ways:

(1) Substitution

$$x + 3y = 11$$
$$5x - 2y = 4$$

From the first equation $x = 11 - 3y$.
Substituting in the second equation gives:

$$5(11 - 3y) - 2y = 4$$
$$55 - 15y - 2y = 4$$
$$17y = 51$$
$$y = 3$$

So, $x = 11 - (3 \times 3)$
$$x = 2$$

(2) Graphical

$$x + 3y = 11$$
$$5x - 2y = 4$$

Draw the graphs of the two relationships.

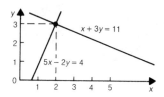

The point of intersection of the lines (2, 3) gives the solution of the equations.

(3) Addition/subtraction of equations

$x + 3y = 11$
$5x - 2y = 4$

multiply the first equation by 5:

$5x + 15y = 55$

then subtract the second equation:

$17y = 51$
$y = 3$

now substitute to find x:

$x + 3 \times 3 = 11$
$x = 2$

(4) Matrices

The equations:

$$x + 3y = 11$$
$$5x - 2y = 4$$

can be written:

$$\begin{pmatrix} 1 & 3 \\ 5 & -2 \end{pmatrix} \begin{pmatrix} x \\ y \end{pmatrix} = \begin{pmatrix} 11 \\ 4 \end{pmatrix}$$

Applying the inverse matrix:

$$\frac{1}{17} \begin{pmatrix} 2 & 3 \\ 5 & -1 \end{pmatrix}$$

to each side of this matrix equation, we have:

$$\begin{pmatrix} 1 & 0 \\ 0 & 1 \end{pmatrix} \begin{pmatrix} x \\ y \end{pmatrix} = \frac{1}{17} \begin{pmatrix} 2 & 3 \\ 5 & -1 \end{pmatrix} \begin{pmatrix} 11 \\ 4 \end{pmatrix}$$

$$\Rightarrow \begin{pmatrix} x \\ y \end{pmatrix} = \begin{pmatrix} 2 \\ 3 \end{pmatrix}$$

Simultaneous equations arise from a variety of different physical situations, and the study of their solution forms an important branch of algebra. The idea of a **matrix** was developed through the study of equations of this type.

Sine A trigonometric function. In a right-angled triangle the sine of an angle is the ratio:

$$\frac{\text{opposite side}}{\text{hypotenuse}}$$

212

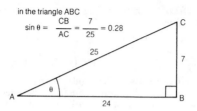

in the triangle ABC

$$\sin \theta = \frac{CB}{AC} = \frac{7}{25} = 0.28$$

The value of θ can now be determined by examination of a table of values of the sine function $\theta = 16.3°$.

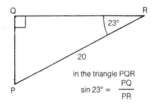

in the triangle PQR

$$\sin 23° = \frac{PQ}{PR}$$

But from the table of values of the sine function $\sin 23° = 0.391$. Hence:

$$0.391 = \frac{PQ}{20}$$

and $PQ = 7.82$.

213

Singular A matrix is singular if it has no inverse. For example:

$$\begin{pmatrix} 3 & 6 \\ 2 & 4 \end{pmatrix} \quad \text{or} \quad \begin{pmatrix} 2 & 1 & 3 \\ 1 & 4 & 5 \\ 4 & 9 & 13 \end{pmatrix} \quad \text{or} \quad \begin{pmatrix} 1 & 2 & 3 \\ 0 & 0 & 0 \\ 4 & 5 & 6 \end{pmatrix}$$

The **determinant** of a singular matrix is zero, and there is a simple linear relationship between the **rows** and **columns** of the matrix.

Singular matrices are associated with sets of linear equations without unique solutions. For example:

matrix $\begin{pmatrix} 3 & 6 \\ 2 & 4 \end{pmatrix}$ $\begin{aligned} 3x + 6y &= 9 \\ 2x + 4y &= 6 \end{aligned}$ $\begin{aligned} 3x + 6y &= 1 \\ 2x + 4y &= 7 \end{aligned}$

                              system with   system with
                              many solutions   no solutions

Skew lines Skew lines are non-parallel, non-intersecting lines.

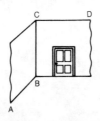

214

In the diagram above, the line AB marking the boundary between floor and wall is skew to the line CD between wall and ceiling.

Skew distribution Data which when represented graphically produces a non-**symmetrical** profile.

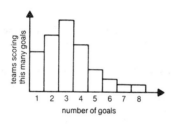

The bar chart above represents the frequencies of various numbers of goals scored by soccer teams in a league.

The distribution is skew, the frequencies do not group around a central average.

Slant

The distance l from the vertex of a right-cone to a point on the edge of the circular base, measured along the curved surface of the cone, is called the slant height of the cone.

This measurement is used in calculating the area of the curved surface of a right cone. Area $= \pi r l$

Slide rule A calculating aid looking similar to a simple rule, but containing a sliding *logarithmic* scale, which can be used to multiply, divide, extract roots, etc.

Slide rules were extensively used by engineers, draughtsmen and others involved in doing large numbers of calculations.

It has been largely replaced by the electronic calculator as a practical calculating aid.

Small circle A circle drawn on the surface of a sphere, whose centre does not coincide with the centre of the sphere.

The term is used most commonly with reference to the geometry of the earth. Lines of **latitude** are small circles (apart from the **equator**, which is a **great circle**) whose centres lie along a common diameter of the earth.

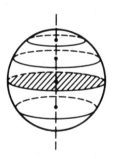

Solid of revolution The solid formed by rotating a plane surface area through 360° about an axis.

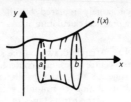

Pictured above is the solid formed by rotating the area under the curve $y = f(x)$ between $x = a$ and $x = b$ about the x-axis.

The rotated area is:

$$\int_a^b y\,dx$$

The volume of the solid of revolution is:

$$\int_a^b \pi y^2\,dx$$

Solution The process and result of solving a mathematical problem often in equation form. Examples:

The solution to $3x + 4 = x + 12$ is $x = 4$.

The solution to $dy/dx = x$ is $y = x^2/2 + A$.

Speed The speed of a body is the rate of distance travelled per unit of time.

Average speeds are calculated over a time interval by:

$$\frac{\text{distance travelled}}{\text{time taken}}$$

For example, an inter-city train travels two hundred miles in 2½ hours. Average speed:

$$\frac{200}{2\frac{1}{2}} = 80 \text{ miles per hour.}$$

The instantaneous speed of a body at a particular time is the **limit** of the average speeds over a **sequence** of time intervals that approach zero.

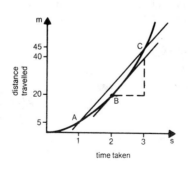

219

The graph above shows the distance travelled by a body in free fall.

The average speed over the second and third seconds is measured by the **gradient** of the **chord** AC. Average speed:

$$\frac{45 - 5}{3 - 1} = 20 \text{ ms}^{-1}$$

The instantaneous speed after two seconds is measured by the gradient of the **tangent** at B. Speed:

$$\frac{40 - 20}{1} = 20 \text{ ms}^{-1}$$

Sphere A circular solid formed by the set of points in space all equidistant from a given fixed point.

A football, a soap bubble and a globe are all examples of roughly spherical shapes.

The fixed point is called the centre of the sphere, and the common distance of its surface points from the centre is called the radius.

The **cartesian** equation of a sphere with centre at (a, b, c) and radius r is:

$$(x - a)^2 + (y - b)^2 + (z - c)^2 = r^2$$

The surface area of a sphere of radius r is given by the formula:

$$A = 4\pi r^2$$

The volume enclosed by a sphere of radius r is given by the formula:

$$V = \frac{4}{3}\pi r^3$$

Spiral A general term given to plane curves or curved surfaces which resemble characteristic patterns found in the natural world.

sea shell

snail shell

spring

Examples are the **logarithmic** or equiangular spiral.

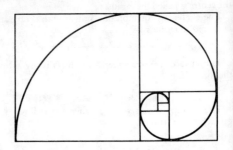

This spiral can be constructed by successive dissections of a rectangle into squares, and the inscribing of **quadrants** of circles.

Spread A term used to denote the dispersion of data from a measure of **average**. Two commonly used measures of spread are **interquartile range** and **standard deviation**. Examples:

Data A   19, 21, 15, 25, 20, 22, 18, 20, 17, 23
Data B   20, 0, 40, 30, 10, 5, 35, 2, 38, 20

The means of these two sets of data are both 20. However the differences of the individual items of data from the means are:

For A   1, 1, 5, 5, 0, 2, 0, 3, 3
For B   0, 20, 20, 10, 10, 15, 15, 18, 18, 0

The mean difference from the mean for A is
26/10 = 2.6.
The mean difference from the mean for B is
126/10 = 12.6.

Measured in this way the data in set B has a greater
spread away from the mean.

Square   1) A square is a **quadrilateral** with four
equal sides, whose interior angles are all 90°.

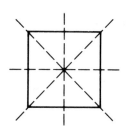

A square is **symmetrical** about its diagonals and the
perpendicular bisectors of its sides, which are all
concurrent.
A square has rotational symmetry of **order** 4.

2) To square a quantity is to multiply that quantity by itself. For example, 5 squared (written 5^2) $= 5 \times 5 = 25$.

3) If 81 is the square of 9, then 9 is the square root of 81 (written $\sqrt{81}$).

Positive real numbers have two square roots:

$$\sqrt{6.25} = + 2.5 \text{ or } - 2.5.$$

4) Square centimetre (cm^2), square metre (m^2), square kilometre (km^2) are all units used in measuring **area**.

A surface with area 5 cm^2, has the same area as five squares of side 1 cm.

Standard deviation A commonly used measure of the **spread** of individual items of **data** from the **mean** of the set of items. For example, the heights in centimetres of ten small plants are:

6.3, 5.9, 5.2, 7.1, 6.6, 6.8, 7.4, 5.3, 5.8, 5.6
The mean height is $62/10 = 6.2$ cm.

The deviations of each of the individual heights from the mean are:

$+0.1, -0.3, -1.0, +0.9, +0.4, +0.6, +1.2, -0.9$
$-0.4, -0.6$

The squares of these deviations are:

0.01, 0.09, 1.0, 0.81, 0.16, 0.36, 1.44, 0.81, 0.16
0.36

224

The mean of these squares of the deviations is:

$$\frac{5.2}{10} = 0.52.$$

The **square root** of this mean is 0.721.

This quantity, the root, mean, square, deviation from the mean is called the standard deviation.

The formula for calculating standard deviation is:

$$\sqrt{\frac{\sum\limits_{1}^{n} (x_i - \bar{x})^2}{n}}$$

for n items of data $x_1, x_2 \ldots$ with a mean of $\bar{x}$

Standard index form   A method of expressing numbers, popular in science and engineering, in which numbers are written in the form:

A $\times$ 10^n

Where $1 \leqslant A < 10$ and n is an integer.

For example:

$$734.6 = 7.346 \times 10^2$$
$$0.0063 = 6.3 \times 10^{-3}$$

This notation is particularly useful for dealing with very large or small numbers.

Stationary point   A function $y = f(x)$ has a stationary point at (a, b) if the **derivative** of $f(x)$ is zero at $x = a$, i.e., $f'(a) = 0$.

$f(x) = x^3 - 6x^2 + 9x$

The gradient of the graph of a function is zero at stationary points. Three types can be identified:

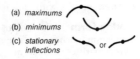

(a) *maximums*

(b) *minimums*

(c) *stationary inflections*

Statistics The study of methods of analysing large quantities of data.

The measures of the various properties of sets of data, for example, **means**, deviations, etc., are also known as statistics.

Structure The underlying properties of a mathematical system are called its structure. Much of modern algebra is devoted to a study of the underlying structures of various systems. For example:

the set of numbers {1, 2, 3, 4} together with the operation of multiplication modulo 5 exhibits a particular structure known as **group structure**. Other systems may exhibit the same group structure, and thus comparisons between the two systems may be made.

Subgroup A **group** H whose elements form a **subset** of a group G, and that shares the same operation with G is called a subgroup of G. For example:

The set {0, 1, 2, 3} under addition modulo 4 is a group

+	0	1	2	3
0	0	1	2	3
1	1	2	3	0
2	2	3	0	1
3	3	0	1	2

The subset {0, 2} under the same operation forms a subgroup

+	0	2
0	0	2
2	2	0

Subset If a **set** B comprises elements that are themselves members of another set A, B is called a subset of A (written B⊂A or A⊃B). All sets are subsets of a **universal set** $\mathscr{E}$. The **empty set** ∅ is a subset of every set.

If A contains elements not in B then B is a **proper** subset of A.

Substitution In mathematics, the replacement of one expression by another in order to simplify is known as substitution.

In particular used as a method of simplifying certain integrals. For example:

$$\int \frac{1}{\sqrt{1 - x^2}} \, dx \quad \text{substitute } x = \sin\theta$$

$$\int \frac{1}{\sqrt{1 - \sin^2\theta}} \cos\theta \, d\theta =$$

$$\int 1 \, . d\theta = \theta + K$$
$$= \sin^{-1} x + K$$

Subtend An angle is subtended by an arc or line when the angle 'stands upon' the arc or line.

In the diagram below the arc AB subtends the angle Ø at the circumference.

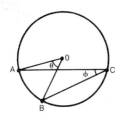

Subtraction One of the four basic operations of arithmetic. It is related to situations where one quantity is removed from another or when the difference between two quantities is to be calculated. Subtraction is the **inverse** operation to **addition**.

Sufficient Considering the statement 'if x is 3 and y is 4; then $x + y$ is 7', of the two conditions:

1) 'x is 3 and y is 4'
2) '$x + y$ is 7'

(1) is a sufficient condition of (2), i.e., if (1) is true, then (2) follows.

However (2) is not a sufficient condition of (1), since if (2) is true (1) need not follow.

Sum The result of adding two or more quantities is called the sum of the quantities.

Supplementary Two angles whose sum is 180° are said to be supplementary.

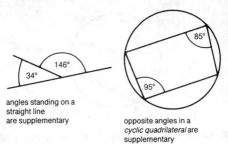

angles standing on a
straight line
are supplementary

opposite angles in a
cyclic quadrilateral are
supplementary

Surd A surd is an expression involving unresolved roots of numbers. Examples:

$$\sqrt{2} \quad \sqrt{5} + \sqrt{7} \quad \sqrt[3]{8} \quad \sqrt{3} - \sqrt{2} \quad \text{etc.}$$

Some expressions involving surds can be simplified by various techniques. For example:

$$\sqrt{8} = \sqrt{4 \times 2} = \sqrt{4} \times \sqrt{2} = 2\sqrt{2}$$

Symbol A sign used to denote a quantity, operation, relation or other mathematical entity. Examples:

$$+ \quad \{\} \quad \vee \quad \int \quad = \quad x$$

Symmetry Plane figures may exhibit two types of symmetry.

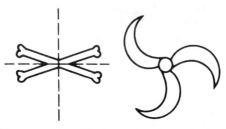

Line symmetry The figure above is equally distributed about the two dotted lines of symmetry.

It has line symmetry of order 2.

Rotational symmetry The figure above maps onto itself under rotations of 120°, 240° and 360° about the centre point.

It has rotational symmetry of order 3.

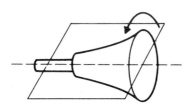

Solid shapes may have symmetry about a plane, or about an axis of rotation.

Tangent 1) A trigonometric function. In a right-angled triangle, the tangent of an angle is the ratio:

$$\frac{\text{opposite side}}{\text{adjacent side}}$$

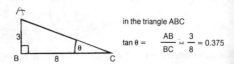

in the triangle ABC

$$\tan \theta = \frac{AB}{BC} = \frac{3}{8} = 0.375$$

From a table of values of the tangent function the value of θ can be found: θ = 20.6°.

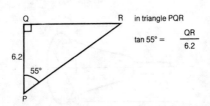

in triangle PQR

$$\tan 55° = \frac{QR}{6.2}$$

But from a table of values of the tangent function tan 55° = 1.43.

So $\dfrac{QR}{6.2} = 1.43$ and QR = 6.2 × 1.43 = 8.87

2) A tangent to a curve at a point is a straight line that touches the curve at that point, and has the same **gradient** as the curve at that point.

Terminating   A terminating decimal is one that has a finite number of decimal digits. For example, 0.41, 0.825, 0.1629385794.

All terminating decimals are equivalent to fractions having **denominators** whose **prime factors** are from the set {2, 5}. Examples:

$$\frac{3}{40} \qquad \frac{7}{125} \qquad \frac{13}{2000} \qquad \text{etc.}$$

Ternary A number system based on grouping in threes. Ternary notation uses the symbols 0, 1 and 2, and the place values are:

	...	81	27	9	3	1

So, sixty is 2 0 2 0 in ternary.

Tessellation A regular pattern of tiles covering a surface is called a tessellation.

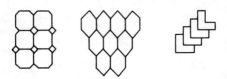

Tetrahedron A tetrahedron is a **polyhedron** with four triangular faces. A triangular based pyramid.

Theorem A significant general conclusion based on certain assumptions, which can be applied to various particular problems is called a theorem. For example, **Pythagoras Theorem**.

Tonne A tonne is a unit of mass equivalent to 1000 Kg.

this Land Rover has a mass of about one tonne

Topology The study of certain geometrical properties of figures that remain unchanged under particular types of transformations.

For simple plane networks properties such as the number and types of **nodes**, the number of **arcs**, etc., are topological properties which remain unchanged under the transformations considered.

Torus A torus is a doughnut shaped surface.

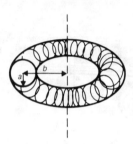

It is the result of rotating in space a circle of radius about an axis, in the plane of the circle, at a distance from the centre of the circle.

The surface area $= 4\pi^2 ab$
The volume $= 2\pi^2 a^2 b$

Trajectory The path of a moving body, ofte. used in the context of bodies moving under th. influence of the earth's (or other large object's gravitational field.

Transcendental Irrational numbers such as π and **e**, which cannot be obtained as the result of solving a **polynomial** equation with **rational coefficients** are called transcendental numbers.

Other numbers, including irrationals, which can result from solving such equations are called algebraic.

$x^2 = 2$ has $\sqrt{2}$ as a solution. Hence $\sqrt{2}$ is algebraic.

Transformation A change in spatial or algebraic work involving the **mapping** of one figure or space to another, or one expression or form to another, often by use of a **substitution**.

The study of various types of transformations is central to work in both algebra and geometry.

Transitive A relation R between elements of a set has the transitive property if it follows that when any three elements x, y and z are related xRy and yRz then xRz. For example:

The relation > ('is greater than') defined on the set of real numbers is transitive: $7 > 6$ and $6 > 4$ implies that $7 > 4$.

The relation ↑ ('is half of') defined on the set of real numbers is not transitive: $3 ↑ 6$ and $6 ↑ 12$ does not imply $3 ↑ 12$.

Translation A geometrical transformation of the plane or of space of the form: $x \rightarrow x + a$, $y \rightarrow y + b$, $z \rightarrow z + c$, etc.

A translation is equivalent to referring to a set of new axes which are parallel to the original set.

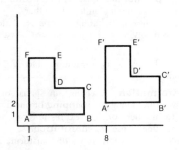

Translations are often written in **vector** form, in this case:

$$\begin{pmatrix} x \\ y \end{pmatrix} \rightarrow \begin{pmatrix} x \\ y \end{pmatrix} + \begin{pmatrix} 7 \\ 1 \end{pmatrix}$$

238

The image of a shape after a translation is **congruent** to the original shape.

Transpose The transpose M' of a **matrix** M is obtained by interchanging the **rows** and **columns** of the matrix.

$$M = \begin{pmatrix} 3 & 1 \\ 2 & 4 \\ 6 & 5 \end{pmatrix} M' = \begin{pmatrix} 3 & 2 & 6 \\ 1 & 4 & 5 \end{pmatrix}$$

If a matrix holds information concerning a relationship, then the transposed matrix will hold corresponding information concerning the **inverse** relationship.

Transversal The name given to a line that intersects with a given set of lines.

transversal

Trapezium A trapezium is a quadrilateral with one pair of **parallel** sides.

If the parallel sides are of lengths a and b, and are separated by a distance h, the area of the trapezium is equal to:

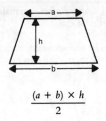

$$\frac{(a + b) \times h}{2}$$

Traversability A network that can be traced with one continuous stroke of a pen, without retracing an **arc** twice is called traversable.

traversable

not
traversable

Tree diagram A diagram used to represent compound events and to enable their **probabilities** to be evaluated.

The tree diagram above represents the eight possible outcomes from tossing a fair coin three times.

The probabilities of individual events are marked on the appropriate branches of the tree, and the

240

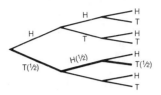

probability of a compound event such as getting a tail, followed by head followed by tail, can be calculated by multiplying the probabilities along the branches.

Probability T, H, T = ½ × ½ × ½ = ⅛.

Trial A trial is the repetition of a particular experiment under controlled conditions in a statistical investigation.

Triangle A triangle is a three-sided polygon. Triangles can be classified in a number of ways.

a scalene triangle has no two sides equal

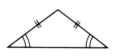

an isosceles triangle has two equal sides (and thus two equal angles)

241

an equilateral triangle
has three equal sides
(and thus three equal
angles)

a right-angled triangle
contains an interior
angle equal to a right
angle

Trigonometry A branch of mathematics concerned, at its simplest level, with the measurement of triangles. Unknown angles or lengths are calculated by using trigonometrical **ratios** such as **sine**, **cosine** and **tangent**.

Trigonometry also involves the study of the properties of the various trigonometrical functions, and their applications to the solutions of problems in many branches of mathematics.

Trinomial Any **polynomial** with just three terms is called a trinomial. For example, $x^4 + x^2 - 5$ or $3x^2 + 4x + 1$.

Truncated A truncated cone or prism is produced from the original solid by removing portions with cutting planes as in the diagrams below.

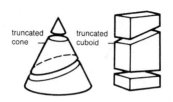

truncated cone   truncated cuboid

Turning point A point on a graph, of a function, which is a local **maximum** or **minimum** value of the function.

At a turning point the value of the **derivative** of the function is zero.

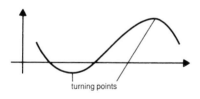

turning points

Unbounded A function is unbounded over an **interval** if a value of the function can be found within the interval to exceed any given number however big.

243

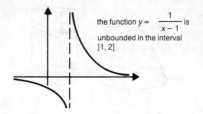

the function $y = \dfrac{1}{x-1}$ is unbounded in the interval $]1, 2]$

Unconditional An inequality such as $x - 1 < x$, which is true for all values of x, is called an unconditional inequality.

Unicursal A curve traced in one sweep, that starts and finishes at the same point, and in which no part of the curve is retraced is called a unicursal curve.

Union A **set** operation. The union of two sets A and B is the set of elements contained in A together with the elements contained in B. For example:

If A = {1, 2, 3, 4} and B = {3, 4, 5, 6}
The union of A and B is {1, 2, 3, 4, 5, 6}

The symbol ∪ is used to represent the opperation.
Hence, A ∪ B = {1, 2, 3, 4, 5, 6}.

Units The standard measures of mass, length, time, etc., are called units. For example, the **metre** is a unit of length, the **kilogram** is a unit of mass.

The term unit is also used to denote 1 or unity. For example, a 'circle of unit radius' is a circle of radius 1 of whatever unit is being employed.

Unity The term unity is used to mean 1.

Universal set The universal set, which is represented by the symbol ℰ, is the set from which all other sets being considered in a problem are considered to be drawn from. All sets are thus **subsets** of the universal set ℰ.

Upper bound An upper bound of a set of numbers is a number which no member of the set exceeds.

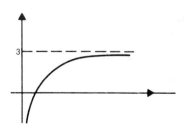

For the set of values of the function:

$$f(x) = \frac{3x - 1}{x}$$

with **domain** the set of positive real numbers, an upper bound is 3.

Value The absolute value, or **modulus**, of a real number is the positive numerical value of the number, ignoring any negative signs. The symbol for absolute value is | |. For example:

$$| -9.3| = 9.3 \qquad |7.2| = 7.2$$

The value of a **polynomial** for a particular number, is the numerical result of evaluating the polynomial with that number substituted in place of the **variable**. For example, the value of $4x^3 - 3x^2 + 7$ at $x = -2$ is -37.

Variable A quantity that can take on a range of values is called a variable. For a **function** $y = f(x)$, x is called the **independent** variable and can take on any value represented in the **domain** of the function. y is called the dependent variable and takes on values represented in the **range** of the function.

For example, for the function $y = \sin x$ the variable x can take on any real number value, whilst y can take on any value in the range $[-1, 1]$.

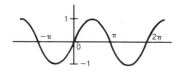

Variance For n items of data x_i with a **mean** of $\bar{x}$, the mean of the squares of the deviations of each x_i from the mean $\bar{x}$ is called the variance.

$$\text{Variance } \sigma^2 = \frac{\sum_{i=1}^{n}(x_i - \bar{x})^2}{n}$$

The square root of the variance $\sqrt{\sigma^2}$ is known as the **standard deviation**.

Vector A quantity that has properties of both **magnitude** and direction is called a vector quantity. Forces, velocities, displacements, etc., are vectors. The characteristic properties of vectors have been abstracted to produce a theory of vectors, which forms an important branch of pure and applied mathematics.

Vectors are often represented by directed line segments like **AB**, whose length and direction represent magnitude and direction. Vectors add up

247

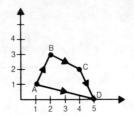

around vector polygons $\mathbf{AB} + \mathbf{BC} + \mathbf{CD} = \mathbf{AD}$.
Vectors are often expressed in **component** form
relative to some standard basis,

$$\mathbf{AB} = \begin{pmatrix} 1 \\ 2 \end{pmatrix} \text{ relative to the unit vectors along the}$$

x and y axis.

Velocity Velocity is a measure of **speed** in a
particular direction. It is the rate of change of
displacement of a body per unit time. Velocity is a
vector quantity.

Venn diagram A diagram showing the relationships between **sets**, by representing them as
regions enclosed by **simple closed curves**.

This Venn diagram represents the relationship between the sets:

A = {1, 2, 3, 4, 5, 6} B = {4, 5, 6, 7, 8}
C = {5, 6}

Vertex In a **polygon** the point of intersection of
two adjacent sides is called a vertex.

In a **polyhedron** it is the point of intersection of
two edges.

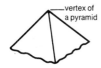
vertex of
a pyramid

vertex of
a triangle

Vertical The direction **perpendicular** to the
plane of the horizon is the vertical. In mathematics
the term vertical is often used to denote measurements perpendicular to some defined base line. For
example, the vertical height of a triangle.

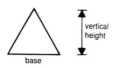
vertical
height

base

Volume The volume of a solid is a measure of the amount of space occupied by the solid.

The volumes of certain simple solids can be easily calculated.

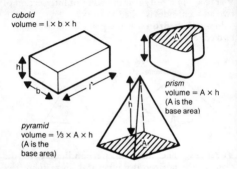

cuboid
volume = l × b × h

prism
volume = A × h
(A is the base area)

pyramid
volume = ⅓ × A × h
(A is the base area)

Vulgar fraction A simple fraction with **integer numerator** and **denominator** is called a vulgar fraction. For example

$$\frac{3}{4}, \quad \frac{13}{19}, \quad \frac{721}{859}.$$

Yard An imperial unit of length equal to three feet. There are 1760 yards in one mile.

Zero The **cardinal** number associated with the **empty set**. Zero is represented by the numeral 0 and is the identity element for addition: $x + 0 = x$ for any real number x.

Zone A zone is part of a sphere bounded by two parallel cutting planes.

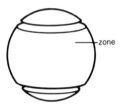

The surface area of a zone of a sphere of radius r, produced by cutting planes distance h apart is:

$$2\pi r h$$